SpringerBriefs in Electrical and Computer Engineering

Speech Technology

Series Editor

Amy Neustein

For further volumes:
http://www.springer.com/series/10043

Editor's Note

The authors of this series have been hand selected. They comprise some of the most outstanding scientists—drawn from academia and private industry—whose research is marked by its novelty, applicability, and practicality in providing broad-based speech solutions. The Springer Briefs in Speech Technology series provides the latest findings in speech technology gleaned from comprehensive literature reviews and *empirical investigations* that are performed in both laboratory and *real life* settings. Some of the topics covered in this series include the presentation of real life commercial deployment of spoken dialog systems, contemporary methods of speech parameterization, developments in information security for automated speech, forensic speaker recognition, use of sophisticated speech analytics in call centers, and an exploration of new methods of soft computing for improving human–computer interaction. Those in academia, the private sector, the self service industry, law enforcement, and government intelligence are among the principal audience for this series, which is designed to serve as an important and essential reference guide for speech developers, system designers, speech engineers, linguists, and others. In particular, a major audience of readers will consist of researchers and technical experts in the automated call center industry where speech processing is a key component to the functioning of customer care contact centers.

Amy Neustein, Ph.D., serves as editor in chief of the *International Journal of Speech Technology* (Springer). She edited the recently published book *Advances in Speech Recognition: Mobile Environments, Call Centers and Clinics* (Springer 2010), and serves as quest columnist on speech processing for Womensenews. Dr. Neustein is the founder and CEO of Linguistic Technology Systems, a NJ-based think tank for intelligent design of advanced natural language-based emotion detection software to improve human response in monitoring recorded conversations of terror suspects and helpline calls.

Dr. Neustein's work appears in the peer review literature and in industry and mass media publications. Her academic books, which cover a range of political, social, and legal topics, have been cited in the Chronicles of Higher Education and have won her a pro Humanitate Literary Award. She serves on the visiting faculty of the National Judicial College and as a plenary speaker at conferences in artificial intelligence and computing. Dr. Neustein is a member of MIR (machine intelligence research) Labs, which does advanced work in computer technology to assist underdeveloped countries in improving their ability to cope with famine, disease/illness, and political and social affliction. She is a founding member of the New York City Speech Processing Consortium, a newly formed group of NY-based companies, publishing houses, and researchers dedicated to advancing speech technology research and development.

Raghunath S. Holambe
Mangesh S. Deshpande

Advances in Non-Linear Modeling for Speech Processing

 Springer

Raghunath S. Holambe
Department of Instrumentation
SGGS Institute of Engineering
    and Technology
Vishnupuri
Nanded 431606
India

Mangesh S. Deshpande
Department of E & TC Engineering
SRES College of Engineering
Kopargaon 423603
India

ISSN 2191-8112
ISBN 978-1-4614-1504-6
DOI 10.1007/978-1-4614-1505-3
Springer New York Heidelberg Dordrecht London

e-ISSN 2191-8120
e-ISBN 978-1-4614-1505-3

Library of Congress Control Number: 2012931407

Printed on acid-free paper

Springer is part of Springer Science+Business Media (www.springer.com)

To our families
For their love, encouragement and support

Preface

Speech production and perception, man's most widely used means of communication, has been the subject of research and intense study for more than 10 decades. Conventional theories of speech production are based on linearization of pressure and volume velocity relations and the speech production system is modeled as a linear source-filter model. This source-filter model is the foundation of many speech processing applications such as speech coding, speech synthesis, speech recognition and speaker recognition technology. However, this modeling technique neglects some nonlinear aspects of speech production. The main purpose of this book is to investigate advanced topics in nonlinear estimation and modeling techniques and their applications to speaker recognition.

The text consists of six chapters that are outlined in detail in Chap. 1. Chapter 2 reviews the fundamentals of speech production and speech perception mechanisms. Some important aspects of physical modeling of speech production system like vocal fold oscillations, the turbulent sound source, aerodynamics observations regarding nonlinear interactions between the air flow and the acoustic field etc. are discussed in this chapter. In Chap. 3, the linear as well as nonlinear modeling techniques of the speech production system are discussed. The widely used source-filter model, its limitations and introduction to dynamic system model are covered in this chapter. Finally, different parametric as well as nonparametric approaches for approximations of nonlinear model are presented.

Advanced topics in nonlinear estimation and modeling are investigated in Chap. 4. Introduction to Teager energy operator (TEO), energy separation algorithms and noise suppression capability of TEO is discussed in this chapter. In Chap. 5, the speech production process is modeled as an AM-FM model which overcomes the limitations of linear source-filter model of speech production and features derived from it like linear prediction cepstral coefficients (LPCC) and mel frequency cepstral coefficients (MFCC). Demodulation techniques like energy separation algorithm using TEO and Hilbert transform demodulation are discussed in this chapter. Based on the foundational Chaps. 2–5, in Chap. 6, an application of the nonlinear modeling techniques is discussed. This chapter covers the performance evaluation of different features based on nonlinear modeling techniques applicable to a speaker identification system. Session variability is one of the challenging tasks in speaker identification. This variability in terms of mismatched environments seriously degrades the identification performance. In order to address the problem of environment mismatch due to noise, different types of robust features are discussed in this chapter. These features make use of nonlinear aspects of speech production model and outperform the most widely accepted MFCC features. The proposed features like Teager energy operator based cepstral coefficients (TEOCC) and amplitude-frequency modulation (AM-FM) based 'Q' features show significant improvement in speaker identification rate in mismatched environments. The performance of these features is evaluated for different types of noise signals in the NOISEX-92 database with clean training and noisy testing environments.

More recently, speech and signal processing researchers have espoused the importance of nonlinear techniques. As this book covers the basics as well as some applications related to speaker recognition technology, this book may be very useful for the researchers working in the speaker recognition area. As compared to the state-of-the-art features which are based on speech production or speech perception mechanism, a new idea is explored to combine the speech production and speech perception systems to derive robust features.

Acknowledgments

We are grateful to many teachers, colleagues and researchers, who directly or indirectly helped us in preparing this book. Very special cordial thanks go to Dr. Hemant A. Patil, Professor, Dhirubhai Ambani Institute of Information and Communication Technology (DA-IICT), Gujarat, for his thorough comments and astute suggestions which immensely enhanced the quality of the work. We are thankful to Dr. S. R. Kajale, Director, Shri Guru Gobind Singhji Institute of Engineering and Technology, Nanded and Dr. D. N. Kyatanavar, Principal, Prof. R. P. Sood, Director, SRES's College of Engineering, Kopargaon, for their motivation and constant support while preparing the manuscript. We are also thankful to all faculty members of the Department of Instrumentation Engineering, Shri Guru Gobind Singhji Institute of Engineering and Technology, Nanded as well as Department of Electronics and Telecommunication Engineering, SRES's College of Engineering, Kopargaon, for their direct and indirect help during completion of this work. We would like to acknowledge our colleagues who have involved indirectly with this work, Dr. J. V. Kulkarni, Head, Department of Instrumentation Engineering, VIT, Pune and Dr. D. V. Jadhav, Principal, Bhiwarabai Sawant College of Engineering, Pune. Finally, we wish to acknowledge Dr. Amy Neustein, Editor, Series in Speech Processing (Springer Verlag) for her unusually great help and efforts during the period of preparing the manuscript and producing the book.

Raghunath Holambe
Mangesh Deshpande

Contents

1	**Introduction**	1
	1.1 Linear and Nonlinear Techniques in Speech Processing	1
	1.2 Applications of Nonlinear Speech Processing	2
	1.3 Outline of the Book	6
	1.4 Summary	7
	References	7
2	**Nonlinearity Framework in Speech Processing**	11
	2.1 Introduction	11
	2.2 Nonlinear Techniques in Speech Processing	11
	2.3 Speech Production Mechanism	12
	2.4 Speech Perception Mechanism	14
	2.5 Conventional Speech Synthesis Approaches	16
	2.6 Nonlinearity in Speech Production	17
	2.6.1 Vocal Fold Oscillation	18
	2.6.2 The Turbulent Sound Source	20
	2.6.3 Interaction Phenomenon	21
	2.7 Common Signals of Interest	21
	2.7.1 AM Signals	21
	2.7.2 FM Signals	22
	2.7.3 AM-FM Signals	23
	2.7.4 Discrete Versions	23
	2.8 Summary	23
	References	24
3	**Linear and Dynamic System Model**	27
	3.1 Introduction	27
	3.2 Linear Model	27
	3.3 The Linear Source-Filter Model	30
	3.3.1 Linear Speech Production Model	30

3.3.2   The Vocal Tract Transfer Function   30
            3.3.3   Lossless Tube Modeling Assumptions   32
            3.3.4   Representations Computed from LPC   33
            3.3.5   LP Based Cepstrum.   33
    3.4   Time-Varying Linear Model   33
    3.5   Dynamic System Model   35
    3.6   Time-Varying Dynamic System Model   36
    3.7   Nonlinear Dynamic System Model   36
    3.8   Nonlinear AR Model with Additive Noise   37
            3.8.1   Multi-Layer Perception   38
            3.8.2   Radial Basis Function   39
            3.8.3   Truncated Taylor Series Approximation.   40
            3.8.4   Quasi-Linear Approximation   41
            3.8.5   Piecewise Linear Approximation.   42
    3.9   Summary   42
    References   43

4   Nonlinear Measurement and Modeling Using Teager
    Energy Operator   45
    4.1   Introduction   45
    4.2   Signal Energy   45
    4.3   Teager Energy Operator   46
            4.3.1   Continuous and Discrete Form of Teager
                      Energy Operator   47
    4.4   Energies of Well-Known Signals.   47
            4.4.1   Sinusoidal Signal   47
            4.4.2   Exponential Signal   48
            4.4.3   AM Signal   48
            4.4.4   FM Signal   48
            4.4.5   AM-FM Signal   50
    4.5   Generalization of Teager Energy Operator   50
    4.6   Energy Separation   52
            4.6.1   Energy Separation for Continuous-Time Signals   52
            4.6.2   Energy Separation for Discrete-Time Signals   53
    4.7   Teager Energy Operator in Noise   57
            4.7.1   Noise Suppression Using Teager Energy Operator   57
    4.8   Summary   59
    References   59

5   AM-FM: Modulation and Demodulation Techniques   61
    5.1   Introduction   61
    5.2   Importance of Phase   62
    5.3   AM-FM Model   63
            5.3.1   Amplitude Modulation and Demodulation   64
            5.3.2   Frequency Modulation and Demodulation   65

5.4   Estimation Using the Teager Energy Operator   67
5.5   Estimation Using the Hilbert Transform   68
5.6   Multiband Filtering and Demodulation   69
5.7   Short-Time Estimates   70
      5.7.1   Short-Time Estimate: Frequency   70
      5.7.2   Short-Time Estimate: Bandwidth   72
5.8   Summary   74
References   75

6 Application to Speaker Recognition   77
6.1   Introduction   77
6.2   Speaker Recognition System   78
6.3   Preprocessing of Speech Signal   79
      6.3.1   Pre-Emphasis   79
      6.3.2   Framing   80
      6.3.3   Windowing   81
6.4   Investigating Importance of Different Frequency Bands
      for Speaker Identification   82
      6.4.1   Experiment 1   83
      6.4.2   Experiment 2   83
      6.4.3   Experiment 3   84
      6.4.4   Experiment 4   85
6.5   Speaker Identification Using TEO   86
      6.5.1   Performance Evaluation for Clean Speech   88
      6.5.2   Performance Evaluation for Noisy Speech:
              Speech Corrupted by Car Engine Noise   88
      6.5.3   Performance Evaluation for Noisy Speech:
              Speech Corrupted by Babble Noise   89
      6.5.4   Effect of Feature Vector Dimensions   89
6.6   Speaker Identification Using AM-FM Model Based Features . . .   90
      6.6.1   Set-up1: Combining Instantaneous Frequency
              and Amplitude   90
      6.6.2   Set-up 2: Combining Instantaneous Frequency
              and Bandwidth   93
      6.6.3   Combining Instantaneous Frequency, Bandwidth
              and Post Smoothing   93
6.7   Summary   96
References   96

Index   101

Chapter 1
Introduction

1.1 Linear and Nonlinear Techniques in Speech Processing

For several decades, the traditional approach to speech modeling has been the linear (source-filter) model where the true nonlinear physics of speech production is approximated via the standard assumptions of linear acoustics and one-dimensional plane wave propagation of the sound in the vocal tract. The linear model has been applied with limited success to applications like speech coding, synthesis and recognition. However, to build successful applications, deviations from the linear model are often modeled as second-order effects or error terms. There are strong theoretical and experimental evidences [1–6] for the existence of important nonlinear aerodynamic phenomena during the speech production that cannot be accounted for by the linear model. Thus, the linear model can be viewed only as a first-order approximation to the true speech acoustics which also contains second-order and nonlinear structure [7].

The linear source-filter model assumes that,

- The vocal tract and speech source are uncoupled (thus allowing source-filter separation), however vocal tract and the vocal folds do not function independent of each other, instead there is some form of coupling between them when the glottis is open [5], which results in significant changes in formant characteristics between open and closed glottis cycles [8].
- Airflow through the vocal tract is laminar, however, Teager and Teager [6] have claimed (based on physical measurements) that voiced sounds are characterized by highly complex air flows in the vocal tract involving jets and vortices, rather than well-behaved laminar flow. In addition, the vocal folds will themselves be responsible for further nonlinear behavior, since the muscle and cartilage which comprise the larynx have nonlinear stretching qualities. Such nonlinearities are routinely included in attempts to model the physical process of vocal fold vibration, which have focused on two or more mass models [9–11], in which the movement of the vocal folds is modeled by masses connected by springs with nonlinear coupling.

R. S. Holambe and M. S. Deshpande, *Advances in Non-Linear Modeling for Speech Processing*, SpringerBriefs in Speech Technology, DOI: 10.1007/978-1-4614-1505-3_1, © The Author(s) 2012

In [12], Schoentgen has shown that, the glottal waveform can change its shape at different amplitudes which would not be possible in a strictly linear system where the waveform shape is unaffected by amplitude changes. Models of the glottal pulse also include nonlinearities, for example, the use of nonlinear shaping functions [12, 13] or the inclusion of nonlinear flow [14]. These imply a loss of information which means that the full speech signal dynamics can never be properly captured.

In practical applications, this manifests itself as an increase in the bit rate, less natural speech synthesis and an inferior ability to discriminate speech sounds. The replacement of the linear filter (or parts thereof) with nonlinear operators (models) should enable us to obtain an accurate description of the speech. This in turn may lead to better performance of practical speech processing applications. Even though, some problems are difficult to solve with linear techniques and are more tractable with nonlinear techniques, there are several drawbacks when dealing with nonlinear techniques as mentioned below [15]:

- A lack of a unifying theory between the different nonlinear processing tools (neural networks, homomorphic, polynomial, morphological and ordered statistics filters, etc.).
- The computational burden is usually greater than with the linear techniques.
- It's difficult to analyze a nonlinear system. Several efforts have been taken in this direction, but the results are not general. The most important ones are for nonlinear filters' analysis. For example the time slope transform domain for morphological systems [16–18], obtaining of the root-signal set of median filters [19], that try to emulate the eigenfunctions of a linear system.
- Some times there is not a closed formulation to derive the nonlinear models. This implies that an iterative procedure must be used and local minima problems exist.

1.2 Applications of Nonlinear Speech Processing

Nonlinear speech processing is applicable to almost all the fields of speech processing. Some of the application areas are discussed below:

- **Speech Recognition** The nonlinear model can be useful in phoneme transitions where the vocal tract changes its shape rapidly and conventional analysis methods based on linear models are not appropriate [20, 21]. The AM and FM signals extracted from the nonlinear model can be incorporated as new features into current speech recognition frameworks, like hidden Markov models (HMM). Since the nonlinear speech model is capable of characterizing rapid variations in speech, incorporating such features shows interesting results.

  The nonlinear predictor may be able to perform the task of front end feature extractor and low level classifier simultaneously. A simple recognition system can be designed where each class (which could be of phones, diphones etc.) can be characterized by a nonlinear model. Then, given an input frame of speech, it will be possible to use the sum of the error residual from the predictor over the frame to decide which class the input speech belongs to. Thus the feature extraction and the classification problem are merged together and solved by one unit. As the nonlinear techniques allow to merge feature extraction and classification problem, and to include the dynamics of the speech signal in the model, it is likely to have significant improvements over current methods which are inherently static [15].

- **Speech Coding** In speech coding, it is possible to obtain good results using models based on linear predictive coding, since the residual can be coded with sufficient accuracy, given a high enough bit rate. However, it is also evident that some of the best results in terms of optimizing both quality and bit rate are obtained from codec structures that contain some form of nonlinearity. Analysis-by-synthesis coders fall into this category. For example, in CELP coders the closed loop selection of the vector from the codebook can be seen as a data dependent nonlinear mechanism. It is reasonable to expect that the nonlinear predictor will contribute to improving this long-term prediction and hence the performance of the coder [15].

- **Speech Analysis** The earlier research was more focused on the analysis of read or laboratory speech and written text, therefore the knowledge on running speech perception and speech production mechanisms is still limited. Hence, better fundamental representation of the signal must be formulated. This can be done by analyzing the acoustic signal in a very detailed manner, using corpora based on spontaneous speech. Spontaneous speech is characterized by several styles which reflect the emotional state and the intents of the speaker in communicating a message. However, acoustic features of emotional state have been studied in restricted environments. Therefore, the speech technology is seeking for a new approach to the speech problems, which overcome the problems associated with classical source filter theory and will be able to quantify the nonlinear features of speech time series data and to embed these new features in automatic speech based systems.

- **Speech Synthesis** Speech synthesis technology plays an important role in many aspects of man-machine interaction. The COST action focuses on new techniques for the speech signal generation stage in a speech synthesizer based on concepts from nonlinear dynamical theory. To model the nonlinear dynamics of speech the one-dimensional speech signal is embedded into an appropriate higher dimensional space. This reconstructed state-space representation has approximately the same dynamical properties as the original speech generating system and is thus an effective model for speech synthesis.

To reproduce the natural dynamics of speech, speech models can be operated in the state space domain, such as neural network architectures. The speech synthesized by such methods will be more natural-sounding than linear concatenation techniques because the low dimensional dynamics of the original signal are learnt, which means that phenomena such as inter-pitch jitter are automatically included into the model. In addition to generating high quality speech, other associated tasks will also be addressed. The most important of these is to examine techniques for natural pitch modification which can be linked into the nonlinear model [15].

- **Speech Enhancement** Even though, the progress in the speech processing field is impressive over past decades, there are significant challenges in the systems like recognizers, synthesizers and coders. Actual recognition systems are unable to work with background noise, channel distortion, limited bandwidth, variations of voice quality etc. Synthetic speech is highly intelligible. However, quality is often poor and the resulting signals are characterized by a lack of naturalness and flexibility in terms of changes in voice gender and reflection of emotional states. Contamination of speech signals with background noise reduces the signal-to-noise ratio (SNR). In particular, speech recognition systems experience problems due to noisy environments which is quite acceptable to human listeners. Recently, new nonlinear speech processing methods using artificial neural networks (ANN) have been investigated [22–24] which are shown to be better capable of taking account of non-linearities in the acoustics or electro-acoustic transmission systems [25] and the non-Gaussian nature of speech.

- **Speaker Recognition** For speaker recognition applications, it has been shown that the residual signal of a linear analysis contains enough information to enable human beings for identifying people [26, 27]. It shows that, some speaker-specific information is ignored by a linear analysis. Several researchers have shown that it is possible to improve the identification rates with a combination of linear and nonlinear predictive models.

  Identifying speakers in the background noise is also one of the problems faced by speaker recognition systems. It is shown in [28, 29] that, the use of nonlinear model helps in deriving robust features and hence improves the identification accuracy. In [30], Quatieri extracted speech source and vocal tract parameters based on linear and nonlinear speech production models and applied these techniques to improve speaker recognition in degrading environments characterized by noise and distortion. He shown that the perception-based, non-parametric approach is a more solid foundation than the parametric approach because the parametric approach requires explicit formant, pitch and voicing estimation that is not desirable under degrading conditions.

  Further, for both speech and speaker recognition there is growing experimental evidence by several research groups that using nonlinear aeroacoustic features of the modulation or fractal type as input to HMM-based classifiers (in addition to the standard cepstrum linear features), better recognition performance can be achieved compared with using only linear features. Therefore, work on detecting such features and using them in recognition systems is very promising.

- **Speech Restoration** Speech restoration is another area where the nonlinear model can be useful. The idea is to estimate AM and FM signals in the presence of a noise model that realistically describes the degraded speech signal. Then these estimated signals can be combined to yield the restored speech. This approach is particularly promising when the modulations are extracted using statistical estimation methods [31].

- **Clinical Measurement of Speech** Voice disorders arise due to physiological disease or psychological disorder, accident or surgery affecting the vocal folds and have a profound impact on the lives of patients. This effect is even more extreme when the patients are professional voice users, such as singers, actors, radio and television presenters etc. Commonly used acoustic tools by surgeons and speech therapists, record the changes in acoustic pressure at the lips or inside the vocal tract. These tools [32], can provide potentially objective measures of voice function. Although acoustic examination is only one tool in the complete assessment of voice function, such objective measurement has many practical uses in clinical settings, augmenting the subjective judgment of voice function by clinicians. Actually, normal and disordered sustained vowel speech exhibit a large range of phenomena. This includes nearly periodic or regular vibration (Type I), aperiodic or irregular vibration (Type II) and sounds with no apparent vibration at all (Type III). Titze [33] introduced a typology for these sounds. Normal voices can usually be classified as Type I and sometimes Type II, whereas voice disorders commonly lead to all three types of sounds.

There exists a very large number of approaches to the acoustic measurement of voice function. The most popular of these are the *perturbation* measures as *jitter* and *shimmer* and noise-to-harmonics ratios (NHR) [32]. However, these measurement methods have limitations for the analysis of disordered speech because they make extensive use of extraction of the pitch period or fundamental frequency. Whereas, the concept of pitch period is only valid for Type I sounds and therefore application of these methods based upon periodicity analysis, to any other type of sound is problematic [34]. Furthermore, another reason for the limitations of these methods is that they are based upon classical linear signal processing methods (such as linear prediction analysis and cepstral processing), that are insensitive to the biomechanical nonlinearity and non-Gaussianity in speech. It shows that, there is a clinical need for reliable tools that can characterize all types of disordered voice sounds for a variety of clinical applications.

To address the limitations of classical linear techniques, there has been growing interest in applying tools from nonlinear time series analysis to disordered speech signals in order to characterize and exploit these nonlinear phenomena [32, 35]. Algorithms for calculating the correlation dimension have been applied, which were successful in separating normal from disordered subjects [36]. Correlation dimension and second-order dynamical entropy measures showed statistically significant changes before and after surgical intervention for vocal fold polyps [37]. Instantaneous nonlinear amplitude (AM) and frequency (FM) formant modulations were shown effective at detecting muscle tension dysphonias [38]. For the automated acoustic screening of voice disorders, higher-order statistics lead to improved normal/disordered classification performance when combined with several standard perturbation measures [39].

These studies have shown that, nonlinear time series methods can be valuable tools for the analysis of voice disorders, where they can analyze a much broader range of speech sounds than perturbation measures, and in some cases are found to be more reliable under conditions of high noise.

1.3 Outline of the Book

Speech production and perception, man's most widely used means of communication, has been the subject of research and intense study for more than 10 decades. Conventional theories of speech production are based on linearization of pressure and volume velocity relations and the speech production system is modeled as a linear source-filter model. This source-filter model is the foundation of many speech processing applications such as speech coding, speech synthesis, speech recognition and speaker recognition technology. However, this modeling technique neglects some nonlinear aspects of speech production. The main purpose of this book is to investigate advanced topics in nonlinear estimation and modeling techniques and their applications to speaker recognition.

Chapter 2 reviews the fundamentals of speech production and speech perception mechanisms. Some important aspects of physical modeling of speech production system like vocal fold oscillations, the turbulent sound source, aerodynamics observations regarding nonlinear interactions between the air flow and the acoustic field etc. are discussed.

Chapter 3 describes the linear as well as nonlinear modeling of the speech production system. The widely used source-filter model, its limitations and the dynamic system model are presented over here. Different parametric as well as non-parametric approaches for approximations of nonlinear model are then introduced.

Chapter 4 investigates advanced topics in nonlinear estimation and modeling techniques. The aeroacoustic modeling approach which is discussed in Chap. 2, provides the impetus for using the high resolution Teager energy operator (TEO). Energy separation algorithms for continuous and discrete time signals are then discussed and noise suppression capability of TEO is focused for obtaining the robust features.

The cepstral features like linear prediction cepstral coefficients (LPCC) and mel frequency cepstral coefficients (MFCC) are computed from the magnitude spectrum of the speech frame and the phase spectra is neglected. However, recently many researchers have shows that phase information is also important for audio perception. To overcome the problem of neglecting the phase spectra, the speech production system can be represented as an amplitude modulation-frequency modulation (AM-FM) model. In AM-FM model, the time varying speech signal is represented as a sum of AM and FM signal components. To demodulate the speech signal, to estimation the amplitude envelope and instantaneous frequency components, the energy separation algorithm (ESA) and the Hilbert transform demodulation (HTD) algorithm are discussed in Chap. 5. The framework for AM-FM model is developed in this chapter.

Based on the foundational Chaps. 2–5, Chap. 6 addresses a specific application, the use of nonlinear modeling techniques to develop a speaker identification system.

1.4 Summary

In this chapter, first we discussed the assumptions of the traditional approach to speech modeling i.e., the liner source-filter model and it's limitations. The nonlinear model is a better way of capturing some fine details which could be missed by linear model, however there are several drawbacks of nonlinear techniques also. Finally, different application areas where nonlinear modeling technique can play an important role are highlighted.

References

1. Barney A, Shadle CH, Davies P (1999) Fluid flow in a dynamical mechanical model of the vocal folds and tract: part 1 and 2. J Acoust Soc Am 105(1):444–466
2. Richard G, Sinder D, Duncan H, Lin Q, Flanagan J, Levinson S, Krane M, Slimon S, Davis D (1995) Numerical simulation of fluid flow in the vocal tract. In: Proceedings of Eurospeech, Madrid
3. McGowan RS (1988) An aeroacoustics approach to phonation. J Acoust Soc Am 83(2): 696–704
4. Thomas TJ (1986) A finite element model of fluid flow in the vocal tract. Comput Speech Lang 1:131–151
5. Kaiser JF (1983) Some observations on vocal tract operation from a fluid flow point of view. In: Titze IR, Scherer RC (eds) Vocal fold physiology: biomechanics, acoustics, and phonatory control. Denver Center for the Performing Arts, Denver, pp 358–386
6. Teager HM, Teager SM (1989) Evidence for nonlinear sound production mechanisms in the vocal tract. In: Hardcastle W, Marchal A (eds) Speech production and speech modeling, vol 55. NATO Advanced Study Institute Series D, Bonas
7. Mclaughlin S, Maragos P (2007) Nonlinear methods for speech analysis and synthesis. In: Marshall S, Sicuranza GL (eds) Advances in nonlinear signal and image processing. Hindawi Publishing Corporation, New York
8. Brookes DM, Naylor PA (1988) Speech production modelling with variable glottal reflection coefficient. In: Proeedings of IEEE international conference on acoustics, speech, and signal processing (ICASSP'88), vol 1, pp 671–674, New York, NY
9. Steinecke I, Herzel H (1995) Bifurcations in an asymmetric vocal-fold model. J Acoust Soc Am 97(3):1874–1884
10. Ishizaka K, Flanagan JL (1972) Synthesis of voiced sounds from a two-mass model of the vocal chords. Bell Syst Tech J 51(6):1233–1268
11. Koizumi T, Taniguchi S, Hiromitsu S (1987) Two-mass models of the vocal cords for natural sounding voice synthesis. J Acoust Soc Am 82(4):1179–1192
12. Schoentgen J (1990) Non-linear signal representation and its application to the modelling of the glottal waveform. Speech Commun 9(3):189–201
13. Schoentgen J (1992) Glottal waveform synthesis with volterra shaping functions. Speech Commun 11(6):499–512
14. Hegerl GC, Hoge H (1991) Numerical simulation of the glottal flow by a model based on the compressible Navier-Stokes equations. In: Proceedings of IEEE international conference on acoustics, speech, and signal processing (ICASSP'91), vol 1, Toronto, ON, pp 477–480
15. Faúndez-Zanuyand M, Kubin G, Kleijn WB, Maragos P, McLaughlin S, Esposito A, Hussain A, Schoentgen J (2002) Nonlinear speech processing: overview and applications. Control Intel Syst 30:1–10

16. Maragos P, Kaiser JF, Quatieri TF (1993) Energy separation in signal modulations with application to speech analysis. IEEE Trans Signal Process 41(10):3024–3051
17. Maragos P (1994) A time slope domain theory of morphological systems: slope transforms and max-min dynamics. In: EUSIPCO'94, pp 971–974
18. Dorst L, Boomgaard R (1993) An analytical theory of mathematical morphology. In: Proceedings international workshop on mathematical morphology and its applications to signal processing, Barcelona
19. Arce G, Gallagher N (1982) State description for the root-signal set of median filters. IEEE Trans Acoust Speech Signal Process 30(6):894–902
20. Nathan KS, Silverman HF (1994) Time-varying feature selection and classification of unvoiced stop consonants. IEEE Trans Speech Audio Process 2(3):395–405
21. Hanson HM, Maragos P, Potamianos A (1993) Finding speech formants and modulations via energy separation: with application to a vocoder. In: Proceedings of IEEE international conference on acoustics, speech, and, signal processing (ICASSP'93), vol 11, pp 716–719
22. Hussain A, Campbell DR (1998) Binaural sub-band adaptive speech enhancement using artificial neural networks. EURASIP J Speech Commun (Special Issue on Robust Speech Recognition for Unknown Communication Channels) 25:177–186
23. Hussain A (1999) Multi-sensor neural network processing of noisy speech. Int J Neural Syst 9(5):467–472
24. Knecht WG, Schenkel ME, Moschytz GS (1995) Neural network filters for speech enhancement. IEEE Trans Speech Audio Proc 3(6):433–438
25. Hienle F, Rabenstein R, Stenger A (1997) Measuring the linear and non-linear properties of electro-acoustic transmission systems. In: Proceedings of international workshop on acoustics, echo and noise cancellation (IWAENC'97), London, pp 33–36
26. Yegnanarayana B, Reddy KS, Kishore SP (2001) Source and system features for speaker recognition using AANN models. In: Proceedings of IEEE international conference on acoustics, speech, and signal processing, Salt Lake city, UT, pp 409–412
27. Yegnanarayana B, Prasanna SRM, Zachariach JM, Gupta SC (2005) Combining evidences from source, suprasegmental and spectral features for a fixed-text speaker verification system. IEEE Trans Speech Audio Process 13(4):575–582
28. Deshpande MS, Holambe RS (2009) Speaker identification based on robust am-fm features. In: Proceedings of second IEEE international conference on emerging trends in engineering and technology (ICETET'09), Nagpur, pp 880–884
29. Deshpande MS, Holambe RS (2011) Robust speaker identification in presence of car noise. Int J Biometrics 3:189–205
30. Quatieri T (2002) Nonlinear auditory modeling as a basis for speaker recognition. Technical report, MIT Lincoln Laboratory, Lexington
31. Lu S, Doerschuk PC (1995) Nonlinear modeling and processing of speech with applications to speech coding. Technical report, Purdue University
32. Baken RJ, Orlikoff RF (2000) Clinical measurement of speech and voice, 2nd edn. Singular Thomson Learning, San Diego
33. Titze IR (1995) Workshop on acoustic voice analysis: summary statement. Technical report, National Center for Voice and Speech, Iowa
34. Godino-Llorente JI, Gomez-Vilda P (2004) Automatic detection of voice impairments by means of short-term cepstral parameters and neural network based detectors. IEEE Trans Biomed Eng 51:380–384
35. Herzel H, Berry D, Titze IR, Saleh M (1994) Analysis of vocal disorders with methods from nonlinear dynamics. J Speech Hear Res 37:1008–1019
36. Zhang Y, Jiang JJ, Biazzo L, Jorgensen M (2005) Perturbation and nonlinear dynamic analyses of voices from patients with unilateral laryngeal paralysis. J Voice 19:519–528
37. Zhang Y, McGilligan C, Zhou L, Vigand M, Jiang JJ (2004) Nonlinear dynamic analysis of voices before and after surgical excision of vocal polyps. J Acoust Soc Am 115:2270–2277 (2004)

38. Hansen JHL, Ceballos GC, Kaiser JF (1998) A nonlinear operator-based speech feature analy-
    sis method with application to vocal fold pathology assessment. IEEE Trans Biomed Eng
    45(3):300–313
39. Alonso J, de Leon J, Alonso I, Ferrer M (2001) Automatic detection of pathologies in the voice
    by hos based parameters. EURASIP J Appl Signal Process 4:275–284

Chapter 2
Nonlinearity Framework in Speech Processing

2.1 Introduction

This chapter presents a survey of nonlinear methods for speech processing. Recent developments in nonlinear science have already found their way into a wide range of engineering disciplines, including digital signal processing. It is also important and challenging to develop the nonlinear framework for speech processing because of the well known nonlinearities in the human speech production mechanism.

2.2 Nonlinear Techniques in Speech Processing

The use of nonlinear techniques in speech processing is a rapidly growing area of research. There are large variety of methods found in the literature, including linearization as in the field of adaptive filtering, introduced by Haykin [1] and various forms of oscillators and nonlinear predictors, as introduced by Kubin [2]. Nonlinear predictors are part of the more general class of nonlinear autoregressive models. Various approximations for nonlinear autoregressive models have been proposed in two main categories: parametric and nonparametric methods. In [3], Kumar et al. show how parametric methods are exemplified by polynomial approximation, locally linear models and state dependent models. Another important group of parametric methods is based on neural nets, radial basis functions approximations, as demonstrated by Birgmeier [4, 5], de Maria and Figueiras [6], and Mann and McLoughlin [7], multi-layer perceptrons as shown by Tishby [8], Wu et al. [9] and Thyssen et al. [10] and recurrent neural nets, as seen in the work of Wu et al. [9] and Hussain [11]. Nonparametric methods include various nearest neighbor methods [12] and kernel-density estimates.

Another class of nonlinear speech processing methods include models and digital signal processing algorithms proposed to analyze nonlinear phenomena of the fluid dynamics type in the speech airflow during speech production as proposed by Teager

R. S. Holambe and M. S. Deshpande, *Advances in Non-Linear Modeling for Speech Processing*, SpringerBriefs in Speech Technology, DOI: 10.1007/978-1-4614-1505-3_2, © The Author(s) 2012

[13]. The investigation of the speech airflow nonlinearities can result in development of nonlinear signal processing systems suitable to extract related information of such phenomena. Recent work by Maragos et al., includes speech resonances modeling using AM-FM model [14]. Further, measuring the degree of turbulence in speech sounds using fractals is explained by Maragos and Potamianos in [15]. The nonlinear speech features are applied to the problem of speech recognition by Dimitriadis et al. in [16], to speech vocoders by Maragos et al. [15] and Potamianos et al. [17]. We have also applied it to the problem of speaker recognition [18, 19]. To understand various linear and nonlinear techniques used for speech processing, it is very essential to know about speech production and perception mechanisms.

2.3 Speech Production Mechanism

Speech is generated as one exhales air from the lungs while the articulators move. Thus speech sound production is a filtering process in which a speech sound source excites the vocal tract filter. The source either is periodic, causing voiced speech, or is noisy (aperiodic), causing unvoiced speech. The source of the periodicity for the former is found in the larynx, where vibrating vocal cords interrupt the airflow from the lungs, producing pulses of air. The lungs provide the airflow and pressure source for speech and the vocal cords usually modulate the airflow to create many sound variations. However, it is the vocal tract that is the most important system component in human speech production. Figure 2.1 shows the anatomy of the speech production system. The vocal tract is a tube-like passageway made up of muscles and other tissues and enables the production of different sounds. For most of the sounds, the vocal tract modifies the temporal and spectral distribution of power in the sound waves, which are initiated in the glottis.

After leaving the larynx, air from the lungs passes through the pharyngeal and oral cavities, then exits at the lips. For nasal sounds, air is allowed to enter the nasal cavity (by lowering the velum), at the boundary between the pharyngeal and oral cavities. The velum (or soft palate) is kept in a raised position for most speech sounds, blocking the nasal cavity from receiving air. During nasal sounds, as well as during normal breathing, the velum lowers to allow air through the nostrils. In the vocal tract, the tongue, the lower teeth and the lips undergo significant movements during speech production.

Figure 2.2 shows the simplified model of the vocal tract with side branches [20]. The vocal tract anatomically divides into four segments: the hypopharyngeal cavities, the mesopharynx, the oral cavity and the oral vestibule (lip tube). The hypopharyngeal part of the vocal tract consists of the supraglottic laryngeal cavity and the bilateral conical cavities of the *piriform fossa*. The mesopharynx extends from the aryepiglottic fold to the anterior palatal arch. The oral cavity is the segment from the anterior palatal arch to the incisors. The oral vestibule extends from the incisors to the lip opening [20]. In the nasal cavity, there are a number of paranasal cavities that contribute anti-resonances (zeros) to the transfer function of the vocal tract [21] and has

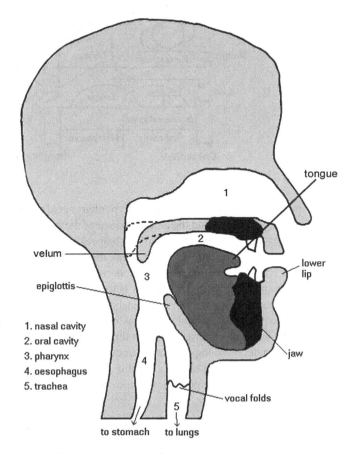

Fig. 2.1 Anatomy of human speech production system

no movable structures. Its large interior surface area significantly attenuates speech signal energy. The opening between the nasal and pharyngeal cavities controls the amount of acoustic coupling between the cavities and hence the amount of energy leaving the nostrils. Since the nasal cavity has a complicated structure and quite large individual differences, it also provides a lot of speaker-specific information. The piriform fossa is the entrance of the esophagus and is shaped like twin cone-like cavities on the left and right sides of the larynx. Because of its obscure form and function, the piriform fossa has usually been neglected in many speech production models. However, introducing the piriform fossa module into the production model causes spectral structure changes in frequency region between 4 kHz and 5 kHz, which can fit the real acoustic speech spectrum well. In addition, the piriform fossa cavities are speaker dependent and less changed during speech production. Dang and Honda suggested that, piriform fossa should be regarded as one important 'cue' for finding speaker-specific features [22]. Further they have tried to obtain such infor-

Fig. 2.2 Simplified model of vocal tract [20]

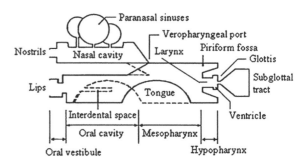

mation using MRI measurements and noted that, the hypopharyngeal resonance, i.e., the resonance of the laryngeal cavity and the antiresonance of the piriform fossa, are more stable than other formants among vowels of each speaker, while they vary to a greater extent from speaker to speaker [23, 24]. Thus the hypopharyngeal cavity also plays an important role to determine individual characteristics.

The most important aspect of speech production is the specification of different phones via the filtering actions of the vocal tract described in terms of its resonances, called formants, owing to poles in the vocal tract transfer function. The formants are often abbreviated F_i like F_1 means the formant with the lowest frequency. In voiced phones, the formants often decrease in power as a function of frequency (due to the general low pass nature of the glottal excitation); thus F_1 is usually the strongest formant. For some phones, inverse singularities of the vocal tract transfer function (zeros) exist and cause anti-resonances, where the speech power dips much more than usual between formants.

2.4 Speech Perception Mechanism

In the past, several studies have been aimed at identifying perceptual cues used by listeners, i.e., how human listener's auditory system processes speech sounds? The discipline of sound perception in general is referred to as psychoacoustics. Techniques adopted from psychoacoustics are extensively used in audio and speech processing systems for reducing the amount of perceptually irrelevant data [25].

Studies by Pickles, of the human hearing mechanism show that the processing of speech and other signals in the auditory system begins with a frequency analysis performed in the *cochlea* [26]. In the human peripheral auditory system, the input stimulus is split into several frequency bands within which two frequencies are not distinguishable. The ear averages the energies of the frequencies within each *critical band* and thus forms a compressed representation of the original stimulus. This observation has given impetus for designing perceptually motivated filter banks as front-ends for speech and speaker recognition systems.

Psychoacoustics studies have shown that human perception of the frequency content of sounds, either for pure tones or for speech signals, does not follow a linear scale. This research has led to the idea of defining subjective pitch of pure tones. Thus for each tone with an actual frequency, f, measured in Hz, a subjective pitch is measured on a scale called the *mel* scale. As a reference point, the pitch of 1 kHz tone, 40 db above the perceptual hearing threshold, is defined as 1000 mels. The subjective pitch in mels increases less and less rapidly as the stimulus frequency is increased linearly. The subjective pitch is essentially linear with the logarithmic frequency beyond about 1000 Hz.

Another important subjective criterion of the frequency contents of a signal is the *critical band* that refers to the bandwidth at which subjective responses, such as loudness, becomes significantly different. The loudness of a band of noise at a constant sound pressure remains constant as the noise bandwidth increases up to the width of the critical band; after that increased loudness is perceived. Similarly, a subcritical bandwidth complex sound (multitone) of constant intensity is about as loud as an equally intense pure tone of a frequency lying at the center of the band, regardless of the overall frequency separation of the multiple tones. When the separation exceeds the critical bandwidth, the complex sound is perceived as becoming louder. It shows the existence of an auditory filter in the vicinity of the tone that effectively blocks extraneous information from interfering with the detection of the tone. This vicinity is called a critical band and can be viewed as the bandwidth of each auditory filter. It is known that the width of the critical band increases with the higher frequency of the tone being masked [27]. The Bark scale is a good approximation to psychoacoustic critical band measurement.

More recently, majority of the speech and speaker recognition systems have converged to the use of feature vectors derived from a filter bank that has been designed according to some model of the auditory system. There are number of forms used for these filters, but all of them are based on a frequency scale that is roughly linear below 1 kHz and roughly logarithmic above this point. Some of the widely used frequency scales include the MEL scale [28], the BARK scale [28, 29] and the ERB (Equivalent Rectangular Bandwidth) scale [30]. In general, the peripheral auditory system can be modeled as a bank of bandpass filters, of approximately constant bandwidth at low frequencies and of a bandwidth that increases in rough proportion to frequency at higher frequencies. The popular Mel frequency cepstral coefficients (MFCCs) incorporate the MEL scale, which is represented by the following equation (since it is based on human experimental data, there are a number of approximations and models that have been used.):

$$F_{\text{Mel}} = 2595 \log_{10}\left(1 + \frac{F_{\text{Hz}}}{700}\right) \tag{2.1}$$

where F_{Hz} denotes the real frequency, and F_{Mel} denotes the perceived frequency. The Mel scale is approximately linear up to 1000 Hz and logarithmic thereafter. Another well-known mapping is the Bark-scale [28, 29]. For the Bark scale, several analytical formulae have been proposed. One of them is the one proposed by Zwicker and Terhardt [31]:

$$F_{\text{Bark}} = 13 \tan^{-1}\left(0.76\,\frac{F_{\text{Hz}}}{1000}\right) + 3.5 \tan^{-1}\left(\frac{F_{\text{Hz}}}{7500}\right)^2 \tag{2.2}$$

Another example of Bark-scale approximation is as following:

$$F_{\text{Bark}} = 6\,\sinh^{-1}\left(\frac{F_{\text{Hz}}}{600}\right) \tag{2.3}$$

At the low end of the Bark scale (<1000 Hz), the bandwidths of the critical band filters are found to be about 100 Hz and in higher frequencies the bandwidths reach up to about 3000 Hz [32].

Moore and Glasberg proposed the ERB scale modifying Zwickers loudness model [30]. The ERB scale is a measure that gives an approximation to the bandwidth of filters in human hearing using rectangular bandpass filters. There are several different approximations of the ERB scale exist. The following is one of such approximations

$$\text{ERB} = 21.4\,\log_{10}\left(1 + 4.37\,\frac{F_{\text{Hz}}}{1000}\right) \tag{2.4}$$

Above discussion shows that, the human ear processes fundamental frequency on a logarithmic scale rather than a linear scale. Therefore, the auditory frequency analysis is most frequently modeled by a bank of bandpass filters whose bandwidths increase with increasing frequency.

2.5 Conventional Speech Synthesis Approaches

Conventional methods of speech synthesis are discussed by Mclaughlin and Maragosin in [33]. Conventionally, the approaches to speech synthesis depend on the type of modeling used. This may be a model of the speech organs themselves (articulatory synthesis), a model derived from the speech signal (waveform synthesis), or alternatively the use of prerecorded segments extracted from a database and joined together (concatenative synthesis).

Modeling the actual speech organs is an attractive approach, since it can be regarded as being a model of the fundamental level of speech production. An accurate articulatory model would allow all types of speech to be synthesized in a natural manner, without having to make many of the assumptions required by other techniques (such as attempting to separate the source and vocal tract parts out from one signal). However, realistic articulatory synthesis is an extremely complex process and as such, instead of using it in any commercial application it is still used as a research tool.

Waveform synthesizers derive a model from the speech signal as opposed to the speech organs. This approach is derived from the linear source-filter theory of speech production. The resulting quality is extremely poor for voiced speech, sounding very robotic.

Concatenation methods involve joining together prerecorded units of speech which are extracted from a database. The concatenation technique provides the best quality synthesized speech. Although there is a good degree of naturalness in the synthesized output, it is still clearly distinguishable from real human speech.

McAulay and Quatieri developed a speech generation model that is based on a glottal excitation signal made up of a sum of sine waves [34]. Then, they used this model to perform time-scale and pitch modification. Starting with the assumption made in the linear model of speech that the speech waveform $x(t)$ is the output generated by passing an excitation waveform $e(t)$ through a linear filter $h(t)$, the excitation is defined as a sum of sine waves of arbitrary amplitudes, frequencies, and phases. A limitation of all these techniques is that they use the linear model of speech as a basis.

2.6 Nonlinearity in Speech Production

Conventional theories of speech production are based on linearization of pressure and volume velocity relations and it assumed constant within a given cross section of the vocal tract, i.e., a one-dimensional planar wave assumption. We refer to this as the linear source-filter theory. While these approximations have allowed a great deal of progress to be made in understanding how speech sounds are produced and how to analyze, modify, synthesize and recognize sounds, the approximations have led to limitations. In reality, acoustic motion is not the only kind of air motion involved. The air in the vocal tract system is not static, but moves from lungs out of the mouth, carrying the sound field along with it, i.e., it contains a *nonacoustic* component. This nonacoustic phenomena, yielding a difference from the linear source-filter theory and have an impact on the *fine structure* in the speech waveform and thus how speech is processed is explained by Quatieri in [28].

The linear assumption neglects the influence of any nonacoustic motion of the fluid medium. In this model, the output acoustic pressure wave at the lips is due solely to energy from an injection of air mass at the glottis. It is known that, in this process, only a small fraction of the kinetic energy, in the flow at the glottis, is converted to acoustic energy propagated by compression and rarefaction waves [28]. The vocal tract acts as a passive acoustic filter, selectively amplifying some bands while attenuating others.

Fine structure refers to attributes in a speech waveform that can be modeled by rapid variations of parameters of traditional speech models, where rapid means on a time scale of a pitch period. For a source-filter model, the fine structure corresponds to source as well as filter. In a source-filter model, both the spacing and the amplitude of the source glottal pulses during voiced speech are considered fixed. However, these parameters are not fixed and leads to one kind of fine structure. The variation of the fundamental period is called as *jitter* whereas period-to-period change in the pulse amplitude is called as *shimmer*. Another type of fine structure, *diplophonia*, is sometimes seen at the ends of utterances. In diplophonia, every other pitch period is both scaled down in amplitude and shifted in time. These examples of fine structure involve modifying locations and amplitudes of existing pitch pulses. Similarly, rapid variations in the filter are actually caused by interaction between the glottis (source) and the vocal tract. But the source-filter model assumes that the behavior in the glottis is not influenced by effects in the vocal tract and vice-versa. Ananthapadmanabha et al. described a model for such interactions and noted that, the effect of the glottis on the first formant is to modulate both formant frequency and the bandwidth during the open glottal phase and higher formants are less affected [35]. Such modulations are speaker-dependent to the extent that they code differences in detailed glottal behavior, resulting from physiological or other differences between speakers [36].

Even though the acoustic speech waveform and its interpretation in terms of phonetic theory is understood, models which mimic human speech production is not completely understood. Some of the difficulties in this field are related to the inadequacy of a simple source-filter model of speech production where a highly stylized source signal generator drives a slowly time-varying linear filter with negligible interaction between source and filter. Vocal fold oscillation, the turbulent sound source and the interaction phenomena are the important aspects of physical modeling of speech production and nonlinearity plays an important role in all of them. In the following, these three aspects are discussed briefly.

2.6.1 Vocal Fold Oscillation

The vocal folds, together with the aerodynamics associated to the glottis and vocal tract, constitute a self-excited biomechanical oscillator that acts as the sound source during voice production. Under certain instability conditions for the biomechanical parameters such as air pressure, vocal fold tension and glottal area, the air flow through the glottis causes the oscillation, which in turn produces the air pressure wave perceived as voice [37, 38]. Thus, it is a self-excited flow-induced oscillation, which is the same phenomenon that produces the oscillation of buildings by action of the wind, the vibration of airplane wings during flight and the generation of sound in wind musical instruments [39]. This oscillator has a relatively complex dynamical structure, as consequence of nonlinear viscoelastic characteristics of its tissues, collisions between the opposite vocal folds and nonlinear interaction between the airflow and the glottal area. Using mathematical models of that structure, past works by Lucero, [40] and Herzel et al. [41], have shown the existence of several nonlinear phenomena, such as multiple equilibrium positions and limit cycles and several types of bifurcations and chaotic behavior.

Many of the acoustic and perceptual features of an individual's voice are believed to be due to specific characteristics of the quasi-periodic excitation signal provided by the vocal folds. These, in turn, depends on the morphology of the voice organ, the larynx. The anatomy of the larynx is quite complicated and its descriptions may be found in the literature [42]. From an engineering point of view, the larynx is the structure that houses the vocal folds whose vibration provides the periodic excitation. The space between the vocal folds, called the glottis, varies with the motion of the vocal folds and thus modulates the flow of air through them. We now know that the larynx is a self-oscillating acousto-mechanical oscillator. This oscillator is controlled by several groups of tiny muscles housed in the larynx. Some of these muscles control the rest position of the folds, others control their tension and still others control their shape. During breathing and production of fricatives, for example, the folds are pulled apart to allow free flow of air. To produce voiced speech, the vocal folds are brought close together. When brought close enough together, they go into a spontaneous periodic oscillation. These oscillations are driven by Bernoulli pressure (the same mechanism that keeps airplanes aloft) created by the airflow through the glottis. If the opening of the glottis is small enough, the Bernoulli pressure due to the rapid flow of air is large enough to pull the folds toward each other, eventually closing the glottis. This, of course, stops the flow and the laryngeal muscles pull the folds apart. This sequence repeats itself until the folds are pulled far enough away or if the lung pressure becomes too low. Besides the laryngeal muscles, the lung pressure and the acoustic load of the vocal tract also affect the oscillation of the vocal folds. These oscillations are driven from an almost stationary lung pressure. Linear time-invariant systems are unable to produce such oscillations. If we exclude a hypothetical time-varying nervous control input, it results into the conclusion that, the oscillation process is nonlinear. This nonlinearity is routinely included even in simple methods of vocal fold behavior as explained by Flanagan in [43], where it is attributed to nonlinear feedback via the Bernoulli force.

The qualitative changes in the type of steady-state motion of a nonlinear dynamical model such as the vocal folds are referred to as *bifurcations*. They show up as discontinuities when a system parameter is moved across some threshold. For instance, an equilibrium state (e.g. open glottis in unvoiced speech) may bifurcate to a periodic limit-cycle motion (e.g. after a transition to voiced speech). It has become popular to characterize vocal fold models in terms of their bifurcation diagrams [44, 45]. The transition between the model and falsetto registers is also considered a bifurcation [46]. Chaos refers to the steady-state motion of nonlinear dynamical systems characterized by high sensitivity to initial conditions. These motions often appear irregular and initial perturbations diverge exponentially.

Because of nonlinear behavior of vocal fold oscillations, it can be seen that the spectral envelop of the glottal pulse changes with its pitch frequency and the spectral content changes with its amplitude.

2.6.2 The Turbulent Sound Source

Turbulence is the source of noise-like sound in speech. In the linear speech model this has been dealt with by having a white noise source exciting the vocal tract filter. Turbulence is the dominant source for frication, aspiration and whisper, and a partial source in breathy voice and creaky voice. The common picture is that turbulence is a nonlinear phenomenon with strong interaction between the airflow and the acoustic sound field occurring at constrictions and obstacles. Turbulence is one of the prominent examples where both theoretical explanation and experimental evidence for chaos are available [47].

Turbulent airflow shows highly irregular fluctuations of particle velocity and pressure. These fluctuations are audible as broadband noise. Turbulent excitation occurs mainly at two locations in the vocal tract: near the glottis and at constriction(s) between the glottis and the lips. Turbulent excitation at a constriction downstream of the glottis produces fricative sounds or voiced fricatives depending on whether or not voicing is simultaneously present. Measurements and models for turbulent excitation are even more difficult to establish than for the periodic excitation produced by the glottis because, usually, no vibrating surfaces are involved. Because of the lack of a comprehensive model, much confusion exists over the proper sub-classification of fricatives. The simplest model for turbulent excitation is a *nozzle* (narrow orifice) releasing air into free space.

Experimental work has shown that half (or more) of the noise power generated by a jet of air originates within the so-called mixing region that starts at the nozzle outlet and extends as far as a distance four times the diameter of the orifice. The noise source is therefore distributed. Distributed sources of turbulence can be modeled by expanding them in terms of monopoles (i.e., pulsating spheres), dipoles (two pulsating spheres in opposite phase), quadrupoles (two dipoles in opposite phase) and higher-order representations. A much stronger noise source is created when a jet of air hits an obstacle. Depending on the angle between the surface of the obstacle and the direction of flow, the surface roughness and the obstacle geometry, the noise generated can be up to 20 dB higher than that generated by the same jet in free space. Because of the spatially concentrated source, modeling obstacle noise is easier than modeling the noise in a free jet. Experiments reveal that obstacle noise can be approximated by a dipole source located at the obstacle. The above theoretical findings qualitatively explain the observed phenomenon that the fricatives "th" and "f" (and the corresponding voiced "dh" and "v") are weak compared to the fricatives "s" and "sh". The teeth (upper for "s" and lower for "sh") provide the obstacle on which the jet impinges to produce the higher noise levels. A fricative of intermediate strength results from a distributed obstacle when the jet is forced along the roof of the mouth.

Modern theories that attempts to explain turbulence predict the existence of eddies (vortices with a characteristic size λ) at multiple scales [48]. According to the energy cascade theory, energy produced by eddies with large size λ is transferred hierarchically to the small-size eddies which actually dissipate this energy due to viscosity. This multiscale structure of turbulence can in some cases be quantified by *fractals*.

2.6.3 Interaction Phenomenon

The linear, one-dimensional acoustic model is too tightly constrained to accurately model many characteristics of vocal tract. The widely used linear predictive cepstral coefficient (LPCC) and Mel frequency cepstral coefficient (MFCC) features are based on this linear speech production model and assume that the airflow propagates in the vocal tract as a linear plane wave. There is an increasing collection of evidence suggesting that nonaciustic fluid motion can significantly influence the sound field. For example, measurements by Teager reveal the presence of separated flow within the vocal tract [49]. Separated flow occurs when a region of fast moving fluid-a jet-detaches from regions of relatively stagnant fluid. When this occurs, viscous forces (neglected by linear models) create a tendency for the fluid to 'roll up' into rotational fluid structures commonly referred to as *vortices* as shown in Fig. 2.3b. Teager suggested that the presence of traveling vortices, 'smoke rings' could result in additional acoustic sources throughout the vocal tract. This contribution of non-linear excitation sources is something neglected by source-filter theory [50]. Figure 2.3a and b show the linear and nonlinear models of sound propagation along the vocal tract, respectively [28, 51].

Motivated by the measurements of Teager, Kaiser hypothesized that the interaction of the jet flow and the vortices with the vocal tract cavity is responsible for much of the speech fine structure, particularly at high formant frequencies. Then he proposed the need for time frequency analysis methods with greater resolution than short-time Fourier transform (STFT) for measuring fine structure within a glottal cycle. He further argues that the instantaneous formant frequencies may be more important than the absolute spectral shape.

2.7 Common Signals of Interest

Signals that we often use in this book are defined in this section. In Chaps. 4 and 5, we will see how some of the parameters of these signals can be estimated.

2.7.1 AM Signals

An amplitude modulated (AM) signal is the combination of two signals, where one signal is the *carrier*, which is a single frequency sinusoidal signal and the other is the information we want to transmit, i.e., the *baseband* signal. The amplitude modulated

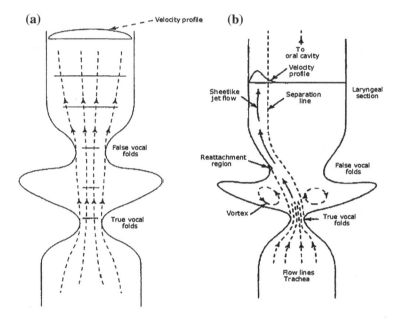

Fig. 2.3 a Linear and **b** nonlinear model of sound propagation along the vocal tract [28, 51]

signal can be modeled as:

$$a(t) = A[1 + km(t)] \tag{2.5}$$

$$s_{AM}(t) = a(t)cos(\Omega_c t) \tag{2.6}$$

where A is the signal amplitude, Ω_c is the carrier frequency (in radians/second), $m(t)$ is the baseband signal and k is the modulation index.

2.7.2 FM Signals

Just like the AM signals, an FM signal is the combination of two signals, where one is the single frequency sinusoidal signal, the carrier and the other is the baseband signal, however, in FM signals the baseband signal is used to change the frequency of the carrier signal.

A frequency modulated signal (FM signal) can be modeled as,

$$\phi(t) = \Omega_c t + \Omega_m \int_0^t q(\tau)d\tau + \theta \tag{2.7}$$

$$s_{FM}(t) = A\,cos(\phi(t)) \tag{2.8}$$

where A is the signal amplitude, Ω_c is the carrier frequency, Ω_m is the maximum frequency deviation with $\Omega_m \in [0, \Omega_c]$, $q(t)$ is the baseband signal with $|q(t)| \le 1$, and θ is the constant phase offset. The instantaneous frequency is defined as the derivative of $\phi(t)$:

$$\Omega_i(t) = \frac{d\phi(t)}{dt} = \Omega_c + \Omega_m q(t) \tag{2.9}$$

2.7.3 AM-FM Signals

The AM-FM signal is the combination of both the AM and FM signals discussed above and it can modeled as,

$$s_{AM-FM}(t) = a(t)cos[\phi(t)]$$

$$= a(t)cos\left(\Omega_c t + \Omega_m \int_0^t q(\tau)d\tau + \theta\right) \tag{2.10}$$

This signal can model the time-varying amplitude and frequency patterns in speech resonances. $s_{AM-FM}(t)$ is a cosine of carrier frequency Ω_c with a time-varying amplitude signal $a(t)$ and a time varying instantaneous frequency signal $\Omega_i(t)$.

2.7.4 Discrete Versions

We can get discrete versions of the AM, FM, and AM-FM signals above by sampling them. We can derive new expressions for these if we substitute t by nT and Ω by ω/T, where ω is the digital frequency (in radians/sample), and T is the sampling period. Finally, the integrations are replaced by sums.

2.8 Summary

This chapter described the speech production and perception mechanisms considering the nonlinearities present in them. Some important aspects of physical modeling of speech production system like vocal fold oscillations, the turbulent sound source, aerodynamics observations regarding nonlinear interactions between the air flow and the acoustic field are discussed in this chapter.

References

1. Haykin S (2001) Adaptive filter theory. Prentice Hall, Upper Saddle River
2. Kubin G (1995) Nonlinear processing of speech. In: Kleijn WB, Paliwal KK (eds) Speech coding and synthesis. Elsevier Science, Amsterdam
3. Kumar A, Mullick SK (1996) Nonlinear dynamical analysis of speech. J Acoust Soc Amer 100:615–629
4. Birgmeier M (1995) A fully Kalman-trained radial basis function network for nonlinear speech modeling. In: Proceedings of IEEE international conference on neural networks (ICNN'95), Perth
5. Birgmeier M (1996) Nonlinear prediction of speech signals using radial basis function networks. In: EUSIPCO'96, vol 1, pp 459–462
6. de Maria FD, Figueiras AR (1995) Nonlinear prediction for speech coding using radial basis functions. In: Proceedings of IEEE international conference on acoustics, speech, and, signal processing (ICASSP'95), pp 788–791
7. Mann I, McLaughlin S (1999) Stable speech synthesis using recurrent radial basis functions. In: Proceedings of European conference on speech communication and technology, vol 5, pp 2315–2318
8. Tishby N (1990) A dynamical systems approach to speech processing. In: Proceedings of IEEE international conference on acoustics, speech, and signal processing (ICASSP'90)
9. Wu L, Niranjan M, Fallside F (1994) Fully vector quantized neural network-based code-excited nonlinear predictive speech coding. IEEE Trans Speech Audio Process 2(4)
10. Thyssen J, Nielsen H, Hansen SD (1994) Non-linear short term prediction in speech coding. In: Proceedings of IEEE international conference on acoustics, speech, and signal processing (ICASSP'94), Australia, pp I-185–I-188
11. Hussain A (1996) Novel artificial neural-network architectures and algorithms for non-linear dynamical system modelling and digital communications applications. Ph.D. thesis, University of Strathclyde, Glasgow
12. Farmer JD, Sidorowich JJ (1988) Exploiting chaos to predict the future and reduce noise. In: Lee YC (ed) Evolution, learning, and cognition. World Scientific, Singapore, pp 277–330
13. Teager HM, Teager SM (1989) Evidence for nonlinear sound production mechanisms in the vocal tract. In: Hardcastle W, Marchal A (eds) Speech production and speech modeling, vol 55. NATO Advanced Study Institute Series D, Bonas, France
14. Maragos P, Kaiser JF, Quatieri TF (1993) Energy separation in signal modulations with application to speech analysis. IEEE Trans Signal Process 41(10):3024–3051
15. Maragos P, Potamianos A (1999) Fractal dimensions of speech sounds: computation and application to automatic speech recogntion. J Acoust Soc Amer 105:1925–1999
16. Dimitriadis D, Maragos P, Potamianos A (2002) Modulation features for speech recognition. In: Proceedings of IEEE international conference on acoustics, speech, and, signal processing (ICASSP'02), pp I-377–I-380
17. Potamianos A, Maragos P (1999) Speech processing applications using an am-fm modulation model. Speech Commun 28:195–209
18. Deshpande MS, Holambe RS (2009) Speaker identification based on robust am-fm features. In: Proceedings of second IEEE international conference on emerging trends in engineering and technology (ICETET'09), Nagpur, pp 880–884
19. Deshpande MS, Holambe RS (2009) Robust q features for speaker identification. In: Proceedings of IEEE international conference on advances in recent technologies in communication and computing (ARTCom'09), Kottayam, Kerala, pp 209–213
20. Honda K (2008) Physiological processes of speech production. Springer, Berlin
21. Kitamura T, Honda K, Takemoto H (2005) Individual variation of the hypopharyngeal cavities and its acoustic effects. Acoust Sci Tech 26:16–26
22. Dang J, Honda K (1997) Acoustic characteristics of the piriform fossa in models and humans. J Acoust Soc Am 101(1):456–465

23. Kitamura T, Takemoto H, Adachi S, Mokhtari P, Honda K (2006) Cyclicity of laryngeal cavity resonance due to vocal fold vibration. J Acoust Soc Am 120(6):2239–2249
24. Dang J, Honda K (1996) An improved vocal tract model of vowel production implementing piriform fossa resonance and transvelar nasal coupling. In: Proceedings of international conference on spoken language processing (ICSLP'96), pp 965–968
25. Kinnunen T (2003) Spectral features for automatic text-independent speaker recognition. Ph.D. thesis, Finland
26. Pickles J (1982) An introduction to the physiology of hearing. Academic Press, London
27. Fletcher H (1940) Auditory patterns. Rev Mod Phys 12:47–65
28. Quatieri TF (2004) Discrete-time speech signal processing. Principles and practice. Pearson Education, London
29. Gold B, Morgan N (2002) Speech and audio signal processing. Wiley, New York
30. Moore BCJ, Glasberg BR (1996) A revision of Zwickers loudness model. Acustica–Acta Acustica 82:335–345
31. Zwicker E, Terhardt E (1980) Analytical expressions for critical band rate and critical bandwidth as a function of frequency. J Acoust Soc Am 68:1523–1525
32. Green DM (1976) An introduction to hearing. Wiley, New York
33. Mclaughlin S, Maragos P Nonlinear methods for speech analysis and synthesis. In: Marshall S, Sicuranza GL (eds) Advances in nonlinear signal and image processing. Hindawi Publishing Corporation, New York
34. McAulay RJ, Quatieri TF (1986) Speech analysis/synthesis based on a sinusoidal representation. IEEE Trans Acoustic Speech Signal Process 34:744–754
35. Ananthapadmanabha TV, Fant D (1982) Calculation of true glottal flow and its components. Speech Commun 1:167–184
36. Jankowski CR (1996) Signal processing using the Teager energy operator and other nonlinear operators. Ph.D. thesis. MIT, Cambridge
37. Titze I (1988) The physics of small-amplitude oscillation of the vocal folds. J Acoust Soc Amer 83:1536–1552
38. Titze I (1994) Principles of voice production. Prentice-Hall, Englewood Cliffs
39. Thompson J, Stewart H (1996) Nonlinear dynamics and chaos. Wiley, New York
40. Lucero J (1999) A theoretical study of the hysteresis phenomenon at vocal fold oscillation onset-offset. J Acoust Soc Amer 105:423–431
41. Herzel H, Knudsen C (1995) Bifurcation in a vocal fold model. Nonlin Dyn 7:53–64
42. Zemlin WR (1968) Speech and hearing science, anatomy, and physiology. Prentice-Hall, Englewood Cliffs
43. Flanagan JL (1872) Speech analysis synthesis and perception, 2nd edn. Springer, New York
44. Lucero J (1993) Dynamics of the two-mass model of the vocal folds: equilibria, bifurcations and oscillation region. J Acoust Soc Amer 94:3104–3111
45. Awrejcewicz J (1991) Bifurcations and chaos in coupled oscillators. World Scientific, Singapore
46. McGowan RS (1993) The quasi-steady approximation in speech production. J Acoust Soc Amer 84:3011–3013
47. Manneville P (1990) Dissipative structures and weak turbulence. Academic Press, Boston
48. Tritton DJ (1988) Physical fluid dynamics, 2nd edn. Oxford University Press, New York
49. Teager HM (1980) Some observations on oral air flow during phonation. IEEE Trans Speech Audio Process 28(5):599–601
50. Hansen JHL, Ceballos GC, Kaiser JF (1998) A nonlinear operator-based speech feature analysis method with application to vocal fold pathology assessment. IEEE Trans Biomed Eng 45(3):300–313
51. Zhou G, Hansen JHL, Kaiser JF (2001) Non-linear feature based classification of speech under stress. IEEE Trans Speech Audio Process 9(3):201–216

Chapter 3
Linear and Dynamic System Model

3.1 Introduction

In this chapter, we have discussed two types of widely used mathematical models, the linear model and the dynamic system model. The linear model and the dynamic system model (often formulated in a specific form called the state-space model) are two distinct types of mathematical structures, both with popular applications in many disciplines of engineering. These two types of the models have very different mathematical properties and computational structures. Instead of using a high-dimensional linear model to represent the temporal correlation structure of a signal, a low-dimensional dynamic system model can be used for efficient computations.

3.2 Linear Model

The linear model is defined as,

$$x[n] = a(x[n-1], x[n-2]...., x[n-N], e[n])$$ (3.1)

where $x[n]$ is the discrete-time signal, $a(.)$ is a general function and $e[n]$ is a white innovation signal. It is referred to as autoregressive (AR) model because the output can be thought of as regressing on itself, i.e., it can be obtained from the feedback of N previous output samples, $x[n-k]$.

The linear AR models are obtained by selecting the function $a(.)$ such that it is linear in both the previous outputs $x[n-k]$ and the innovation signal $e[n]$,

$$x[n] = \sum_{k=1}^{N} a_k x[n-k] + e[n].$$ (3.2)

R. S. Holambe and M. S. Deshpande, *Advances in Non-Linear Modeling*
for Speech Processing, SpringerBriefs in Speech Technology,
DOI: 10.1007/978-1-4614-1505-3_3, © The Author(s) 2012

The coefficients a_k are referred to as the *linear prediction coefficients* and their estimation is termed as *linear prediction analysis*. Quantization of these coefficients, or of a transformed version of these coefficients, is called *linear prediction coding* (LPC) which is useful in speech coding. Here, the linear prediction coefficients are used as linear weights for the past samples and $e[n]$ is additive with unity weight. This model can be used in two modes,

- For *prediction*:

  For voiced speech, $e[n]$ can be thought of as a train of unit samples. Therefore, except for the times at which $e[n]$ is nonzero, i.e., every pitch period, from Eq. 3.2 we can think of $\hat{x}[n]$ as a linear combination of past values of $x[n]$, i.e.,

  $$\hat{x}[n] = \sum_{k=1}^{N} a_k x[n-k], \quad \text{when} \quad e[n] = 0. \tag{3.3}$$

  The system function associated with the Nth order predictor is a finite length impulse response (FIR) filter of length N given as

  $$P(z) = \sum_{k=1}^{N} a_k z^{-k}. \tag{3.4}$$

- For *signal synthesis* or *modeling*:

  Where the prediction residuals are important. The prediction error sequence is given by the difference of the sequence $x[n]$ and its prediction $\hat{x}[n]$, i.e.,

  $$r[n] = x[n] - \hat{x}[n]$$
  $$= x[n] - \sum_{k=1}^{N} a_k x[n-k] \tag{3.5}$$

  and the associated prediction error filter is defined as

  $$A(z) = 1 - \sum_{k=1}^{N} a_k z^{-k}$$
  $$= 1 - P(z). \tag{3.6}$$

  The residual should be white noise. If this assumption is justified, we can replace the actual residual $r[n]$ by a synthetic white noise excitation $\hat{e}[n]$ without loosing speech quality of the synthetic signal $y[n]$ obtained from

$$y[n] = \sum_{k=1}^{N} a_k y[n-k] + \hat{e}[n]. \tag{3.7}$$

If this assumption is not justified, we look for an excitation signal $\hat{e}[n]$ with similar properties as $r[n]$. In speech coding, this is often achieved by indirect waveform matching using analysis-by-synthesis techniques.

The predictor coefficients a_k are determined so that the mean square error is minimized:

$$\min_{(a_1,...,a_k)} \sum_k \left(x[n] - \sum_{k=1}^{N} a_k x[n-k] \right)^2 \tag{3.8}$$

The coefficients are typically obtained using the *Levinson-Durbin algorithm* [1]. Frequency-domain interpretation of the model in Eq. 3.2 is obtained by taking z-transforms of both sides of Eq. 3.2 and solving for the filter transfer function

$$H(z) = \frac{X(z)}{E(z)} = \frac{1}{1 - \sum_{k=1}^{N} a_k z^{-k}}. \tag{3.9}$$

This transfer function has no zeros, and therefore, it is called as the transfer function of an *all-pole* filter. The poles are the roots of the denominator, and they correspond to local maxima in the spectrum. Since a pole represents a local peak in the magnitude spectrum, the model is restricted in the sense that it models only the peaks of the spectrum (resonances of the vocal tract).

Nasal sounds and nasalized vowels include, in addition to the resonances, so-called anti-resonances that result from the closed side-tube formed by the oral cavity. Modeling of these anti-resonances requires zeros in the filter [2, 3]. The resonances of nasal cavity have a large bandwith because viscous losses are high as air flows along its complexely configured surface, quickly damping its impulse response. The closed oral cavity acts as a side branch with its own resonances that change with the place of constriction of the tongue; these resonances absorb acoustic energy and thus are anti-resonances (zeros) of the vocal tract. Therefore, the linear model is not suitable for modeling the anti-resonances or nasal parameters.

This linear model is widely used in speech processing as well as in other areas of signal processing and in control and communication theory. This model can also be defined in the canonical form as discussed below. To define the model in the canonical form, let us first define:

- $o[n] \equiv$ observation data at sample time n, and
- $\mathbf{o} = (o[0], o[1], ..., o[N-1])^T \equiv$ vector of N observation samples.

The canonical form of a linear model can be written as

$$\mathbf{o} = \mathbf{H}\theta + v, \tag{3.10}$$

Fig. 3.1 The source-filter
model of voice production
for voiced and unvoiced
speech [4]

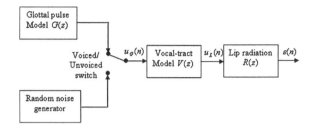

where **o** is the observation or measurement vector, with the dimensionality of $N \times 1$, θ is the parameter vector, with dimensions of $n \times 1$, **H** is the observation matrix, with dimensions of $N \times n$ and v is the observation noise vector, which is of dimensions $N \times 1$. v is a random vector, while **H** and θ are deterministic. The random nature of v makes observation **o** a random vector also.

3.3 The Linear Source-Filter Model

The linear modeling of the speech production was motivated by the lossless tube model of the vocal tract [5]. Fant conducted a study on speech intelligibility [6] and Miller estimated the voice source signal by using the inverse of the first vocal resonance and the vocal fold opening area measured by video [7]. A linear source-tract model was proposed to represent the radiation impedance, vocal tract and the glottal source as linear filters and identified using covariance analysis [8, 9].

3.3.1 Linear Speech Production Model

A general discrete-time linear speech production model, shown in Fig. 3.1, describes the voiced and unvoiced modes of speech separately. Where $s(n)$ is the sampled speech waveform, n is the sample number, $u_L(n)$ is the volume velocity signal at the lips, $u_G(n)$ is the glottal signal (which is the input into the vocal tract), $V(z)$ is the vocal tract transfer function and $R(z)$ is the lip radiation transfer function. The system is assumed to be time-invariant over a time period of about 10–30 ms [10]. The recorded speech, $s(n)$, is a sound pressure wave and is related to the acoustic velocity at the lips, $(u_L(n))$ through the radiation transfer function $(R(z))$. Acoustic theory predicts $R(z)$ to be a first order high-pass filter [8, 9].

3.3.2 The Vocal Tract Transfer Function

The vocal tract filter $V(z)$ can be approximated as an all-pole filter [1, 5, 10]. The all-pole filter characterizing the vocal tract can be derived from the wave-equation

Fig. 3.2 The vocal tract represented by concatenated lossless tubes of equal length but different cross-sectional area [4]

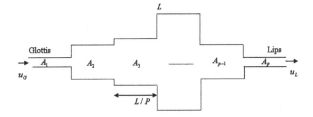

by making approximations and assumptions regarding the acoustic properties of the vocal tract and the behavior of the sound wave in the process. The vocal tract shape is approximated as a concatenation of P rigid tubes each of constant diameter and equal length L/P, where L is the length of the vocal tract (typically 15–17 cm in adults). This is demonstrated in Fig. 3.2. The length of each segment needs to be approximately equal to the distance that sound travels in half a sample period [10, 11].

$$\frac{L}{P} \approx \frac{v_c}{2f_s}, \tag{3.11}$$

where $v_c \approx 20\sqrt{T_v} \approx 350$ m/s is the speed of sound in the vocal tract, $T_v \approx 305$ K is the air temperature in the vocal tract and f_s is the sampling frequency. This means that for a speech signal with $f_s = 16$ kHz sampling frequency, we need $P \approx \frac{2Lf_s}{v_c} \approx 15$ tube segments to approximate the vocal tract.

The objective of this analysis is to derive the transfer function $V(z)$ between the glottal air-flow, $u_G(n)$ and the flow at the lips, $u_L(n)$. The two main assumptions used in this derivation are that there is no energy loss in the vocal tract and that the sound pressure waves in the vocal tract are longitudinal plane waves. The assumption of no energy loss in the vocal tract means that the walls are assumed to be rigid and that there is no turbulent flow or viscosity. The longitudinal plane wave assumption means that the sound pressure wave is independent of cross sectional coordinates.

The analysis of the lossless tube model relies on representing the acoustic wave in the vocal tract as the superposition of the acoustic velocity in the direction of the lips and the acoustic velocity in the direction of the lungs [10]. The flow entering tube, $p + 1$ in the direction of the lips can then be expressed as the scaled and delayed version of the flow entering tube, p depending on the tube's length L/P and the cross-sectional areas of the tubes A_p and A_{p+1}. Similarly, the flow exciting tube p in the direction of the lungs can be expressed in terms of the flow exiting tube $p + 1$. A linear transfer function $V_p(z)$ for one tube-segment can be derived using these expressions and the transfer function for the concatenated tube-segments is derived by multiplying these together, yielding the vocal tract transfer function,

$$V(z) = \frac{z^{-P/2} \prod_{p=0}^{P}(1 + \rho_p)}{1 - \sum_{p=1}^{P} a_p z^{-p}} = \frac{z^{-P/2} G}{1 - \sum_{p=1}^{P} a_p z^{-p}} \tag{3.12}$$

where $G = \prod_{p=0}^{P}(1 + \rho_p)$ is a gain constant, $z^{-P/2}$ is the acoustic delay of the vocal tract and ρ_p the reflection coefficients given as,

$$\rho_p = \frac{A_{p+1} - A_p}{A_{p+1} + A_p} \tag{3.13}$$

and a_p is obtained using Cholesky decomposition or the Levinson-Durbin algorithm [4].

3.3.3 Lossless Tube Modeling Assumptions

The linear model of the vocal tract, $V(z)$, is justified by acoustic theory using the lossless tube model but it is in fact a gross simplification of the complicated effects that occur during voice production. Many assumptions and approximations are made in the derivation of the discrete linear filter model and when the filter parameters of $V(z)$ are estimated. Here we discuss the strengths and weaknesses of the linear filter model of the voice production.

The first step in deriving the transfer function in Eq. 3.12 using the lossless tube model, is to approximate the vocal tract area function as being piecewise constant. By carefully choosing the number of tube segments, P, as explained in Eq. 3.11, any adverse effect of this approximation can be reduced. Using fewer tube segments has an adverse effect on the modeling whereas choosing the number of tube segments above the optimum number only affects the modeling if there is not enough speech data to estimate the parameters.

The acoustic wave in the vocal tract is also assumed not to suffer any energy loss. The walls of the vocal tract are not rigid so acoustic energy is in fact lost due to the vibration of the walls and viscosity and turbulence of the airflow. The mathematical analysis is computationally intractable as it requires the solution to wave equations with difficult boundary conditions and is normally not considered [10]. Usually, this effect is either ignored or accounted for by raising and broadening the formant frequency peaks. The vibration of the walls causes the low frequency formant peaks to broaden and shift. Similarly, the turbulent flow will affect the high frequency formants. The lossless tube model can therefore be adjusted after the estimation process [12, 13].

The acoustic sound wave is assumed to be constant over the cross-section of the vocal tract. This is not a bad approximation if we can assume that the voice source is decoupled from the vocal tract. This has been disputed since the sound pressure wave does not uniformly fill the vocal tract as the vocal folds open [9, 14].

3.3.4 Representations Computed from LPC

Several alternative representations can be derived from the LPC coefficients. If the autocorrelation method is used, the Levinson-Durbin algorithm produces the reflection coefficients ρ_p, $p = 1, ..., P$ as its side-product. They are also called *partial correlation coefficients*. When the vocal tract is modeled with the lossless tube model, at each tube junction, part of the wave is transmitted and the remainder is reflected back [15]. The reflection coefficients are the percentage of the reflection at these discontinuities. Assuming the lossless tube model, ratio of the areas of the adjacent tubes is given by [16]:

$$\frac{A_{p+1}}{A_p} = \frac{1 - \rho_p}{1 + \rho_p} \qquad (3.14)$$

A new parameter set is obtained by taking the logarithm of the area ratio, yielding log area ratios (LAR). Since the LARs are derived from the LP coefficients, they are subject to the assumptions made in LP. To avoid singularity at $|\rho_p| = 1$, an alternative for log area ratios are *arcsin reflection coefficients* [15], simply computed as taking inverse sine of the reflection coefficients. Perceptual linear prediction (PLP) is a form of generalized linear prediction that exploits some of the psychoacoustics principles, including critical band analysis, equal loudness pre-emphasis and the intensity loudness relationship [17].

3.3.5 LP Based Cepstrum

The LPC coefficients are seldom used as features themselves [2]. It is observed by Rosenberg and Sambur that adjacent predictor coefficients are highly correlated [18] and therefore, representations with less correlated features would be more efficient. A popular feature set is linear predictive cepstral coefficients (LPCCs). The relationship between LPC and LPCC was originally derived by Atal in 1974 [13]. From the theoretical point of view, it is comparably easy to convert LP coefficients to LPCC, in the case of minimum-phase signals [1, 9].

3.4 Time-Varying Linear Model

The linear model defined in Eq. 3.2 can be rewritten for the output sequence $o[n]$ as,

$$o[n] = \sum_{k=1}^{K} \phi_k o[n - k] + v[n]. \qquad (3.15)$$

This can be expressed in the canonical form of a linear model with the parameter vector

$$\theta = [\phi_1, \phi_2, ..., \phi_K]^T. \tag{3.16}$$

This is a time-invariant linear model because the parameter vector θ is not a function of time, n. To turn this linear model into a time-varying one, we impose time dependence on θ, resulting in the time-varying LPC model in the form of

$$o[n] = \sum_{k=1}^{K} \phi_{k,n} o[n - k] + v[n]. \tag{3.17}$$

The time dependence of the parameter vector $\phi_{k,n}$ can take a variety of forms. The simplest form is to allow for non-parametric parameter evolution on a block-by-block basis. That is, divide the entire time axis into a sequential set of blocks (or frames) and for each block (consisting of many samples) the LPC parameters $\phi_{k,n}$ become independent of time n. However, the $\phi_{k,n}$ change from one block to the next. Therefore, the time dependency for this type of non-parametric parameter evolution is such that the parameters are piecewise constant over time. The very popular speech coding technique called LPC coding has been based precisely on this type of non-parametric time-varying LPC model [13]. Under this coding scheme, the parameters of the LPC model need to be estimated for each separate block of speech. Such estimated LPC parameters (together with other residual information) are used to represent the original speech samples on a frame-by-frame basis. These estimated parameters have also been used as speech feature vectors for speech recognition.

A more efficient way to characterize the time dependency of the parameter vector θ in the LPC model is to provide a parametric form for the parameter evolution [19]. Then we will have a new set of parameters, Φ, which gives the time-varying function for the LPC parameters over the entire time span, eliminating the need to divide the signal into frames. This type of parametric form of time-varying LPC model can be written as

$$o[n] = \sum_{k=1}^{K} \phi_n(\Phi_k) o[n - k] + v[n], \tag{3.18}$$

where the LPC parameters ϕ_n are explicitly indexed by discrete time index of n and by the new set of parameters Φ_k.

One common parametric form used for time-varying LPC modeling is that of sinusoids. In this case, the parameters Φ that characterize the time-varying nature of the LPC parameters ϕ_n are the amplitudes, angular frequencies and phases of the sinusoids. This sinusoidally modulated time-varying LPC model can be used in very low bit-rate speech coding.

3.5 Dynamic System Model

The dynamic system model has a very different mathematical structure and computational properties from the linear model. Even though it is possible to indefinitely increase the dimensionality of the signal vectors in the linear model so as to achieve the same mathematical description as a dynamic system model, it increases the computations associated with the model. Furthermore, according to the signal generation mechanisms, speech signals are endowed with hidden signal dynamics, which can be represented in much more natural way by dynamic system model rather than its high-dimensional linear model counterpart.

Dynamic system models typically have two essential characteristics. First, the models are statistical in nature, such that the outputs of the systems are characterized not by a deterministic function but by a stochastic process that consists of a rich variety of temporal trajectories. Second, the models are defined with the use of some internal low-dimensional (continuous) state that is hidden from the direct observations. The state summarizes all information at a given time about the past behavior for predicting the statistical properties of the model output in the future. Use of the hidden state permits decoupling of the internal dynamics from the static relationship between the state variable and the observation variable.

The canonical form of the dynamic system model is its state-space formulation. Therefore, the names, a dynamic system model and state-space model are often used synonymously. A generative form of the discrete-time, time-invariant, linear state-space model has the following mathematical representation:

$$\mathbf{x}(k+1) = \mathbf{A}\mathbf{x}(k) + \mathbf{u} + \mathbf{w}(k) \tag{3.19}$$

$$\mathbf{o}(k) = \mathbf{C}\mathbf{x}(k) + \mathbf{v}(k), \tag{3.20}$$

where, $\mathbf{x}(k)$ is a (hidden) state vector at time k, $\mathbf{o}(k)$ is an output (or observation) vector, $\mathbf{w}(k)$ and $\mathbf{v}(k)$ are uncorrelated zero-mean Gaussian noise vectors and $\mathbf{u}$ is a deterministic input vector. Equation 3.19 is often called the state equation and Eq. 3.20 called the observation (or measurement) equation. The state equation represents linear dynamics of the state variable or vector $\mathbf{x}(k)$ using autoregression (i.e., linear prediction). The noise or error term $\mathbf{w}(k)$ represents the degree of inaccuracy in using the linear state equation to describe the true state dynamics. The observation equation Eq. 3.20, on the other hand, is static in nature. It contains no dynamics since the time indices of $\mathbf{o}$ and $\mathbf{x}$ are the same. It represents the noisy relationship between the state vector and the observation vector. Like the noise term in the state equation, the noise term $\mathbf{v}$ in the observation equation represents the degree of inaccuracy in using the linear mapping $\mathbf{o}(k) = \mathbf{C}\mathbf{x}(k)$ to describe the true relationship between the state and observation vectors. Due to the presence of the noise term and due to the possible non-invertibility of matrix $\mathbf{C}$ in the observation equation, the state vector $\mathbf{x}(k)$ cannot be uniquely determined given the observation vector $\mathbf{o}(k)$. In this sense, we say that the state vector in the state-space model is hidden, and the state dynamics described by the state equation Eq. 3.19 is hidden dynamics. This is analogous to hidden Markov model (HMM), where given an observation it is not possible to uniquely determine which (discrete) HMM state is responsible for generating that observation.

3.6 Time-Varying Dynamic System Model

The linear state-space model defined earlier by Eqs. 3.19 and 3.20 is time invariant because the parameters that characterize this model do not change as a function of time k. In this section, the time-invariant linear dynamic system model is extended to its time-varying or nonstationary version, i.e., the parameters of the dynamic system model vary as a function of time. This generalizes Eqs. 3.19 and 3.20 into the following state and observation equations where the parameters are indexed by time frames k and $\acute{k}$:

$$\mathbf{x}(k+1) = \mathbf{A}_k \mathbf{x}(k) + \mathbf{u}_k + \mathbf{w}(k) \tag{3.21}$$

$$\mathbf{o}(k) = \mathbf{C}_{\acute{k}} \mathbf{x}(k) + \mathbf{v}(k). \tag{3.22}$$

Equations 3.21 and 3.22 used different time indices to denote the parameter change over time: k in the state equation Eq. 3.21 and $\acute{k}$ in the observation equation Eq. 3.22 indicates that the evolutions of the parameters in the state equation and those in the observation equation may not have to be synchronous in time.

In the above discussion, the change of parameters over time is typically slower than the change of state $\mathbf{x}(k)$ and of observation $\mathbf{o}(k)$ over time. In other words, the parameter evolution and the state evolution are at distinct time scales. Therefore, we can say that such a model shows hierarchical nonstationarity.

One common technique used for representing the parameter evolution over time is a discrete hidden Markov model (DHMM) [1]. In DHMM, the Markov chain state transition matrix and the levels associated with each state in the Markov chain become the full set of time-varying state-space model parameters. More sophisticated techniques can use the continuous density HMM (CDHMM) to represent the parameter evolution over time in the time-varying state-space model. In this case, the full set of model parameters includes additional variance parameters associated with each HMM state.

3.7 Nonlinear Dynamic System Model

Many physical systems are characterized by nonlinear relationships between various physical variables. Let us consider an example of the speech production process. A dynamic system model can be used to describe the speech production process. The state equation can be used to describe the dynamic articulation process, while the observation equation be used to characterize the relationship between the articulatory variables and the observed speech acoustics. However, it is well known that the relationship between articulation and acoustics in speech is highly nonlinear and that the articulatory dynamics can also be nonlinear [9]. Therefore, it is inadequate to use a linear dynamic system model to represent the articulatory and acoustic processes in speech production.

When the observation equation is nonlinear, the inversion problem of estimating (i.e., making inference about) the hidden state vector from the observation vector sequence becomes much more difficult than the case where the observation equation is linear. One-to-many (static) mappings from an observation vector to a state vector often arise in the observation equation and need to be disambiguated using the information contained in the dynamic state equation. This one-to-many mapping problem is often more serious for the nonlinear observation equation than the linear one with the observation matrix being singular or of less than a full rank [19]. Therefore, while the use of a nonlinear dynamic system model often provides an adequate mathematical abstraction for representing real-world physical processes with hidden dynamics, it is a serious challenge to solve the inversion problem associated with the nonlinear dynamic system model.

The most general form of the time-varying nonlinear dynamic system model, with nonlinearity in both state and observation equations, is as follows in its state-space formulation [20]:

$$\mathbf{x}(k+1) = \mathbf{g}_k[\mathbf{x}(k), \mathbf{u}_k, \mathbf{w}(k)] \tag{3.23}$$

$$\mathbf{o}(\acute{k}) = \mathbf{h}_{\acute{k}}[\mathbf{x}(k), \mathbf{v}(k)]. \tag{3.24}$$

where subscripts k and $\acute{k}$ indicate that the nonlinear functions $\mathbf{g}[.]$ and $\mathbf{h}[.]$ can be time varying and be asynchronous with each other.

Frequently used and a simpler version of the model separates out the noise terms from the nonlinear functions, e.g., the model discussed by Ghahramani et al. in [21]. The time-invariant (stationary) version of such a simplified model has the state-space form of

$$\mathbf{x}(k+1) = \mathbf{g}[\mathbf{x}(k), \mathbf{u}] + \mathbf{w}(k)] \tag{3.25}$$

$$\mathbf{o}(k) = \mathbf{h}[\mathbf{x}(k)] + \mathbf{v}(k). \tag{3.26}$$

3.8 Nonlinear AR Model with Additive Noise

Nonlinear AR models with additive noise are obtained from choosing a function $a(.)$ in Eq. 3.1 which is nonlinear in the previous outputs $x[n-k]$ but linear in the innovation signal $e(n)$ [22, 23].

$$x[n] = a(x[n-1], x[n-2]...., x[n-N]) + e[n]. \tag{3.27}$$

Like the linear model, both the prediction and the synthesis modes can be explored here. For prediction mode, the minimum mean-square error is obtained by the *conditional expectation* of $x[n]$ given the vector of past samples $x[n-1]$ [24, 25].

$$x[n-1] = (x[n-1], x[n-2], ..., x[n-N])^T \qquad (3.28)$$

$$\hat{x}[n] = E(x[n]|x[n-1]) \qquad (3.29)$$

For the practical design of a nonlinear predictor based on Eq. 3.29, various approximations of the conditional expectation operator have been proposed by Casdagli et al. [26] and Farmer et al. [27]. Various parametric approximation methods have been proposed by researchers like, Sicuranza [28], Thyssen et al. [29], Singer et al. [30], Tong [23], Priestley [31] etc. Radial basis function approximations have been proposed by Birgmeier [32, 33] and Diaz et al. [34] whereas Thyssen et al. [29], Lapedes et al. [35], Tishby et al. [36] and Wu et al. [37] have used the multi-layer perceptrons approach. Haykin et al. [38] and Wu et al. [39] have further discussed the recurrent neural net approach.

Several non-parametric methods also play an important role such as, Lorenz's method of analogues [40, 41] which may be the simplest of various nearest neighbor methods discussed by Farmer [27] and Yakowitz [42], which are further extended by Wu [37] as well as Gersho [43] such as nonlinear predictive vector quantization. Another non-parametric approach is based on kernel density estimates of the conditional expectation proposed by Lee et al. [44] and Prakasa Rao [45].

Given such a huge variety of options, design of a good nonlinear predictor for speech signals is a non-trivial issue. With all these choices, it is interesting to note that, practically all the reported studies claim that nonlinear prediction provides approximately 3 dB gain over the linear prediction of running speech. An explanation may be found in [46]. The fact that significant improvements are obtained mostly for voiced speech whereas unvoiced speech is well modeled by the linear component. As about 40% of the running speech is unvoiced, high gain is to be expected for any method during voiced speech. Some of the nonlinear approximations are discussed below. Additional details are discussed by Deng and O'Shaughnessy, and can be found in [19].

3.8.1 Multi-Layer Perception

One common nonlinear function, which can be used either as the static function $\mathbf{g}[.]$ in the state equation or as the function $\mathbf{h}[.]$ in the observation equation, is called the *multi-layer perceptron* (MLP) neural network [47]. The MLP has been shown to possess some strong theoretical properties which sometimes give the MLP an enviable name of *universal nonlinear approximator*.

Let us consider an example of MLP with three layers (input, hidden and output). Let w_{jl} be the MLP weights from input to hidden units and $\mathbf{W}_{ij}$ be the MLP weights from hidden to output units, where l is the input node index, j the hidden node index and i the output node index. Then the output signal at node i can be expressed as a (nonlinear) function $\mathbf{h}(.)$ of all the input nodes (making up the input vector) according to

$$h_i(\mathbf{x}) = \sum_{j=1}^{J} \mathbf{W}_{ij}.s\left(\sum_{l=1}^{L} w_{jl}.x_l\right), \quad 1 \le i \le I \qquad (3.30)$$

where I, J and L are the numbers of nodes at the output, hidden and input layers, respectively. $s(.)$ is the hidden units nonlinear activation function, taken as the standard sigmoid function of

$$s(x) = \frac{1}{1 + exp(-x)}. \qquad (3.31)$$

The derivative of this sigmoid function has the following concise form:

$$\acute{s}(x) = s(x)(1 - s(x)), \qquad (3.32)$$

The parameters that characterize the MLP nonlinearity are the two sets of weights.

3.8.2 Radial Basis Function

Another popular form of nonlinearity is the radial basis function (RBF) neural network proposed by Bishop [47] and Ghahramani et al. [21]. In a vector form, the nonlinearity can be expressed as:

$$\mathbf{h}(x) = \mathbf{W_y} \qquad (3.33)$$

where $\mathbf{W}$ is the weight matrix that consists of the weights connecting the middle layer and the output layer of the RBF network. The (i, j)th element of $\mathbf{W}$ is the connection weight between node i in the output layer and node j in the middle layer. The middle layer consists of the vector y of the components of

$$\mathbf{y} = [y_1(\mathbf{x}), y_2(\mathbf{x}), ..., y_j(\mathbf{x}), ..., y_J(\mathbf{x})]^T, \qquad (3.34)$$

where $y_j(\mathbf{x})$ is the output of the j-th radial basis function (or kernel function) in the middle layer. If the Gaussian function is chosen to be the kernel function,

$$y_j(x) = exp\left\{-\frac{1}{2}(\mathbf{x} - \mathbf{c}_j)^T \sum_j^{-1}(\mathbf{x} - \mathbf{c}_j)\right\}, \qquad (3.35)$$

then the parameters that characterize the RBF network nonlinearity are the weight matrix $\mathbf{W}$, the kernel centers $\mathbf{c}_j$ and the kernel widths $\sum_j (j = 1, 2, ..., J)$.

The kernel function in the RBF should have the desirable localized property. The Gaussian function above satisfies such a property. Another common localized kernel function is

$$y_j(\mathbf{x}) = (\|\mathbf{x}\|^2 + \alpha^2)^{-\alpha}. \qquad (3.36)$$

In a scalar form, each element of the RBF nonlinear function in Eq. 3.33 can be written as

$$h_i(\mathbf{x}) = \sum_{j=1}^{J} \mathbf{W}_{ij}.y_j(\mathbf{x}), \quad 1 \leq i \leq I. \qquad (3.37)$$

3.8.3 Truncated Taylor Series Approximation

The analytical forms of nonlinear functions, such as the MLP and RBF described above, make the associated nonlinear dynamic systems difficult to analyze and make the estimation problems difficult to solve. Approximations are frequently used to gain computational simplifications while sacrificing accuracy for approximating the nonlinear functions.

One very commonly used technique for the approximation is a truncated vector Taylor series expansion. If all the Taylor series terms of order two and higher are truncated, then we have the linear Taylor series approximation that is characterized by the Jacobian matrix $\mathbf{J}$ and the point of Taylor series expansion $\mathbf{x}_0$. For a vector input of dimension n and a vector output of dimension m, this linear approximation can be written as

$$\mathbf{h}(\mathbf{x}) \approx \mathbf{h}(\mathbf{x}_0) + \mathbf{J}(\mathbf{x}_0)(\mathbf{x} - \mathbf{x}_0). \qquad (3.38)$$

Each element of the Jacobian matrix $\mathbf{J}$ is a partial derivative of each vector component of the nonlinear output with respect to each of the input vector components. That is,

$$\mathbf{J}(\mathbf{x}_0) = \frac{\partial \mathbf{h}}{\partial \mathbf{x}_0} = \begin{bmatrix} \frac{\partial h_1(\mathbf{x}_0)}{\partial x_1} & \frac{\partial h_1(\mathbf{x}_0)}{\partial x_2} & \cdots & \frac{\partial h_1(\mathbf{x}_0)}{\partial x_n} \\ \frac{\partial h_2(\mathbf{x}_0)}{\partial x_1} & \frac{\partial h_2(\mathbf{x}_0)}{\partial x_2} & \cdots & \frac{\partial h_2(\mathbf{x}_0)}{\partial x_n} \\ \cdot & & \cdot & \\ \cdot & & \cdot & \\ \cdot & & \cdot & \\ \frac{\partial h_m(\mathbf{x}_0)}{\partial x_1} & \frac{\partial h_m(\mathbf{x}_0)}{\partial x_2} & \cdots & \frac{\partial h_m(\mathbf{x}_0)}{\partial x_n} \end{bmatrix} \qquad (3.39)$$

As an example, for the MLP nonlinearity of Eq. 3.30, the (i, l)th element of the Jacobian matrix is

$$J_{il} = \sum_{j=1}^{J} \mathbf{W}_{ij}.s_j(y).(1 - s_j(y)).w_{jl}, \quad 1 \le i \le I, 1 \le l \le L, \qquad (3.40)$$

where $y = \sum_{i=1}^{L} \mathbf{W}_{jl}x_i$.

As another example, for the RBF nonlinearity Eq. 3.33 or Eq. 3.37 with Gaussian kernels, the (i, l)th element of the Jacobian matrix is

$$J_{il} = \sum_{j=1}^{J} \mathbf{W}_{ij}.\frac{\partial y_j}{\partial x_l}$$

$$= \sum_{j=1}^{J} \mathbf{W}_{ij}.exp\left\{-\frac{1}{2}(\mathbf{x} - \mathbf{c}_j)^T \sum_{j}^{-1}(\mathbf{x} - \mathbf{c}_j)\right\}\left[\sum_{j}^{-1}(\mathbf{x} - \mathbf{c}_j)\right]_l,$$

$$\text{for } 1 \le i \le I, 1 \le l \le L. \qquad (3.41)$$

If the kernel function of Eq. 3.36 is used in the RBF, then the (i, l)th element of the Jacobian matrix becomes

$$J_{il} = -2\sum_{j=1}^{J} \mathbf{W}_{ij}.(\alpha_l + 1)(\|\mathbf{x}\|^2 + \alpha^2)^{-\alpha-1}x_l \quad \text{for } 1 \le i \le I, 1 \le l \le L. \ (3.42)$$

3.8.4 Quasi-Linear Approximation

The linear Taylor series approximation discussed above requires evaluation of the Jacobian matrix in an analytical form. If such a form is not available, or some or all elements of the Jacobian matrix do not exist as in the case where discontinuity exists, then a linear approximation to a nonlinear function can be accomplished by the quasi-linear approximation method, which we discuss now.

Consider a nonlinear vector function $\mathbf{h}(x)$ of a vector random vector $\mathbf{x}$ with its PDF denoted by $p(\mathbf{x})$. The quasi-linear approximation gives

$$\mathbf{h}(\mathbf{x}) \approx E[\mathbf{h}(\mathbf{x})] + \mathbf{A}(\mathbf{x} - E[\mathbf{x}]), \qquad (3.43)$$

where the matrix $\mathbf{A}$ is

$$\mathbf{A} = \left(E[\mathbf{h}(\mathbf{x})\mathbf{x}^T] - E[\mathbf{h}(\mathbf{x})]E[\mathbf{x}^T] \right) \sum^{-1}. \tag{3.44}$$

In the above, $\sum$ is the covariance matrix for $\mathbf{x}$:

$$\sum = E\left[(\mathbf{x} - E[\mathbf{x}])(\mathbf{x} - E[\mathbf{x}])^T \right]. \tag{3.45}$$

Note that in Eq. 3.45, no explicit derivative of the nonlinear function $\mathbf{h}(\mathbf{x})$ is needed in the computation. Rather, some statistics up to the second order need to be computed, using the PDF of $p(\mathbf{x})$ if it is known in advance, or using the estimates of the PDF from empirical data pairs of the input and output of the nonlinear function.

3.8.5 Piecewise Linear Approximation

A very popular technique for simplifying nonlinear functions is *piecewise linear approximation*. This is essentially an extension of the linear Taylor series approximation from one point to multiple points, where these multiple points are determined using some optimality criteria. The essence of this technique is to linearize the (vector-valued) nonlinear function $\mathbf{h}(\mathbf{x})$ about a finite number of points, and to parameterize these linearized mappings by the set of locations of the linearization points $C_x = \{\mathbf{x}_c(i), i = 1, 2, ..., I\}$, together with the vector values of the function at those points $C_h = \{\mathbf{h}(\mathbf{x}_c(i)), i = 1, 2, ..., I\}$ plus the local Jacobian matrices $C_J = \{\mathbf{J}(\mathbf{x}_c(i)), i = 1, 2, ..., I\}$. All these parameters are contained in the form of a *codebook* $C = \{C_x, C_h, C_J\}$, which may be learned from the available data or may be determined from prior knowledge or by computer simulation.

Given the piecewise linearization parameters (i.e., given the codebook), any value of $\mathbf{h}(\mathbf{x})$ can then be approximated by

$$\mathbf{h}(\mathbf{x}) \approx \mathbf{h}(\mathbf{x}_c) + \mathbf{J}(\mathbf{x}_c)(\mathbf{x} - \mathbf{x}_c)$$
$$\mathbf{x}_c = arg \min_{\mathbf{x}_i \in C_x} \| \mathbf{x} - \mathbf{x}_i \| \tag{3.46}$$
$$\mathbf{J}(\mathbf{x}_c) = \frac{\partial \mathbf{h}}{\partial \mathbf{x}} |\mathbf{x} = \mathbf{x}_c,$$

where $\| . \|$ denotes the Euclidean norm.

3.9 Summary

In this chapter, we have discussed two types of mathematical models relevant to speech processing. First, we have discussed the linear model, where in its canonical form the observation vector is linearly related to the observation matrix via the parameter vector. Then the time-varying linear model is presented, where the parameter vector in the general linear model changes with time. The another form of model discussed is the dynamic system model, whose canonical form is called the state-space model. Finally this model is extended to time-invariant and time-varying models and some approximation methods are discussed.

References

1. Rabiner LR, Juang BH (1993) Fundamentals of speech recognition. Prentice-Hall of India, New Delhi
2. Hansen J, Proakis J (2000) Discrete-time processing of speech signals, 2nd edn. IEEE Press, New York
3. Mammone R, Zhang X, Ramachandran R (1996) Robust speaker recognition: a feature based approach. IEEE Signal Process Mag 13:58–71
4. Gudnason J (2007) Voice source cepstrum processing for speaker identification. Ph.D. thesis, University of London
5. Dunn HK (1961) Methods of measuring vowel formant bandwidths. J Acoust Soc Am 33(12):1737–1746
6. Fant G (1960) Acoustic theory of speech production. Mouton, The Hague
7. Miller RL (1959) Nature of the vocal chord wave. J Acoust Soc Am 31:667–677
8. Wong DY, Markel JD, Gray AH (1979) Glottal inverse filtering from the acoustic speech waveform. IEEE Trans Acoust Speech Signal Process 27(4):350–355
9. Quatieri TF (2004) Discrete-time speech signal processing, principles and practice. Pearson Education, Upper Saddle river
10. Rabiner LR, Shafer RW (1989) Digital signal processing of speech signals. Prentice-Hall, Englewood Cliffs
11. Gold B, Morgan N (2002) Speech and audio signal processing. Wiley, New York
12. Makhoul J (1975) Linear prediction: a tutorial review. In. Proceedings of the IEEE, vol 64, pp 561–580
13. Atal BS (1974) Effectiveness of linear prediction characteristics of the speech wave for automatic speaker identification and verification. J Acoust Soc Am 55:1304–1312
14. Teager HM (1980) Some observations on oral air flow during phonation. IEEE Trans Speech Audio Process 28(5):599–601
15. Campbell J (1997) Speaker recognition: a tutorial. Proc IEEE 511(9):1437–1462
16. Honda K (2008) Physiological processes of speech production. Springer, Berlin
17. Hermansky H (1990) Perceptual linear prediction analysis for speech. J Acoust Soc Am 87:1738–1752
18. Rosenberg AE, Sambur MR (1975) New techniques for automatic speaker verification. IEEE Trans Acoust Speech Signal Process 23(2):169–176
19. Deng L, O'Shaughnessy D (2003) Speech processing a dynamic and optimization-oriented approach. Marcel Dekker, New York
20. Tanizaki H (1996) Nonlinear filters—estimation and applications, 2nd edn. Springer, Berlin
21. Ghahramani Z, Roweis S (1999) Learning nonlinear dynamic systems using an em algorithm. Adv Neural Inf Process Syst 11:1–7
22. Segall A (1976) Stochastic processes in estimation theory. IEEE Trans Inf Theory IT-22: 275–286
23. Tong H (1990) Non-linear time series—a dynamical system approach. Oxford University Press, Oxford
24. Papoulis A (1984) Probability, random variables and Stochastic processes. McGraw-Hill, New York

25. Kantner M (1979) Lower bounds for nonlinear prediction error in moving avarage processes. Ann Prob 7(1):128–138
26. Casdagli M, jardins D, Eubank S, Farmer JD, Gibson J, Theiler J, Hunter N (1992) Nonlinear modeling of chaotic time series: theory and applications. In: Kim J, Stringer J (eds) Applied chaos. Wiley, New York, pp 335–380
27. Farmer JD, Sidorowich JJ (1988) Exploiting chaos to predict the future and reduce noise. In: Lee YC (ed) Evolution, learning, and cognition. World Scientific, Singapore, pp 277–330
28. Sicuranza GL (1992) Quadratic filters for signal processing. In: Proceedings of IEEE, vol 80, pp 1263–1285
29. Thyssen J, Nielsen H, Hansen SD (1994) Non-linear short term prediction in speech coding. In: Proceedings of IEEE international conference on acoustics, speech, and signal processing (ICASSP'94), Adelaide, pp I-185–I-188
30. Singer AC, Wornell GW, Oppenheim AV (1994) Nonlinear autoregressive modeling and estimation in the presence of noise. Dig Signal Process 4:207–221
31. Priestley MB (1988) Non-linear and non-stationary time series analysis. Academic Press, London
32. Birgmeier M (1995) A fully kalman-trained radial basis function network for nonlinear speech modeling. In: Proceedings of IEEE international conference on neural networks, (ICNN'95), Perth
33. Birgmeier M (1996) Nonlinear prediction of speech signals using radial basis function networks. In: EUSIPCO'96, vol 1, pp 459–462
34. de Maria FD, Figueiras AR (1995) Radial basis functions for nonlinear prediction of speech in analysis-by-synthesis coders. In: Proceedings of IEEE workshop on nonlinear signal and image processing, Halkidiki
35. Lapedes A, Farber R (1998) How neural nets work. In: Lee YC (ed) Evolution, learning, and cognition. World Scientific, Singapore, pp 231–346
36. Tishby N (1990) A dynamical systems approach to speech processing. In: Proceedings of IEEE international conference on acoustics, speech, and, signal processing (ICASSP'90)
37. Wu L, Niranjan M, Fallside F (1994) Fully vector quantized neural network-based code-excited nonlinear predictive speech coding. IEEE Trans Speech Audio Process 2(4):482–489
38. Haykin S, Li L (1995) Nonlinear adaptive perdiction of nonstationary signals. Signal Process 43:526–535
39. Wu L, Niranjan M (1994) On the design of nonlinear speech predictors with recurrent nets. In: Proceedings of IEEE international conference on acoustics, speech, and signal processing (ICASSP'94), Adelaide, pp II-529–II-532
40. Lorenz EN (1969) Atmospheric predictability as revealed by naturally occurring analogues. J Atmos Sci 26:636–646
41. Bogner RE, Li T (1989) Pattern search prediction of speech. In: Proceedings of IEEE international conference on acoustics, speech, and signal processing (ICASSP'89), Glasgow, pp 180–183
42. Yakowitz S (1987) Nearest neighbor methods for time series analysis. J Time Ser Anal 8(2): 235–247
43. Gersho A (1989) Optimal nonlinear interpolative vector quantization. IEEE Trans Comm 38(9):1285–1287
44. Lee Y, Johnson D (1993) Nonparametric prediction of non-gaussian time series. In: Proceedings of IEEE international conference on acoustics, speech, and signal processing (ICASSP'93), Minneapolis, MN, pp IV-480–IV-483
45. Rao BLSP (1983) Nonparametric functional estimation. Academic Press, Orlando
46. Kubin G (1995) Nonlinear processing of speech. In: Kleijn WB, Paliwal KK (eds) Speech coding and synthesis. Elsevier Science, Amsterdam
47. Bishop C (1997) Neural networks for pattern recognition. Clarendon Press, Oxford

Chapter 4
Nonlinear Measurement and Modeling Using Teager Energy Operator

4.1 Introduction

We begin this chapter by discussing signal energy in general. We then look at an alternative definition, i.e., the Teager energy operator (TEO) and how it can be obtained by considering a second order differential equation, which describes the motion of an object suspended by a spring. This operator is interesting because it has a small time window, making it ideal for local (time) analysis of signals. The analysis of AM–FM signals using the Teager Energy Operator is probably the field where most of the research regarding the operator has been done so far. Energy separation algorithm using TEO is then discussed and finally its noise suppression capability is presented.

4.2 Signal Energy

In electrical systems, the instantaneous power of a system is described as

$$p(t) = |v(t)^2|/R \tag{4.1}$$

or,

$$p(t) = R \times i(t)^2 \tag{4.2}$$

where R, $v(t)$ and $i(t)$ is the resistance, voltage and the current of the system. If we normalize this by choosing $R = 1\Omega$, we see that the power is the square of the input signal, regardless of whether we choose to measure the voltage or the current. We can therefore express the *instantaneous power* as

$$p(t) = |s(t)|^2 \tag{4.3}$$

R. S. Holambe and M. S. Deshpande, *Advances in Non-Linear Modeling for Speech Processing*, SpringerBriefs in Speech Technology, DOI: 10.1007/978-1-4614-1505-3_4, © The Author(s) 2012

where $s(t)$ is either voltage or current.

The energy of the signal is defined as,

$$E = \int_{-\infty}^{\infty} |s(t)|^2 dt \tag{4.4}$$

Note that this is the total energy of the signal and not the instantaneous energy. Looking towards above definition of the energy, the tones at 10 Hz and at 1000 Hz with the same amplitude have the same energy. However, Teager observed that, the energy required to generate the signal at 1000 Hz is much greater than that at 10 Hz. This alternative notion of energy can be understood with the sinusoidal oscillation that occurs with a simple harmonic oscillator given by second order differential equation. By studying the second order differential equation (as mentioned below), we find that the energy to generate a simple sinusoidal signal varies as a function of both amplitude and frequency. Therefore, the above definition seems somewhat odd. This observation is what Kaiser used to derive the Teager Energy Operator (TEO).

4.3 Teager Energy Operator

The second order differential equation for an object with mass m suspended by a spring with force constant k, is given as,

$$\frac{d^2x}{dt^2} + \frac{k}{m}x = 0 \tag{4.5}$$

The solution to above equation is a periodic oscillation given by,

$$x(t) = A\cos(\Omega t + \phi) \tag{4.6}$$

where $x(t)$ is the position of the object at time t, A is the amplitude, Ω is the frequency of the oscillation and ϕ is the initial phase. According to Newtonian physics, the total energy of the object is given as the sum of the potential energy of the spring and the kinetic energy of the object, given by

$$E = \frac{1}{2}kx^2 + \frac{1}{2}m\dot{x}^2 \tag{4.7}$$

By substituting $x(t) = A\cos(\Omega t + \phi)$, we obtain

$$E = \frac{1}{2}mA^2\Omega^2 \tag{4.8}$$

or

$$E \propto A^2\Omega^2 \tag{4.9}$$

From Eq. 4.9, we can see that energy is proportional to both A and Ω. This is the true energy in the harmonic system, i.e., the energy required to generate the signal.

4.3.1 Continuous and Discrete Form of Teager Energy Operator

The simple and elegant form of the Teager energy operator was introduced by Kaiser [1, 2] as

$$\Psi_c[x(t)] = \left[\frac{d}{dt}x(t)\right]^2 - x(t)\frac{d^2}{dt^2}x(t) \tag{4.10}$$

$$\Psi_c[x(t)] = [\dot{x}(t)]^2 - x(t)\ddot{x}(t) \tag{4.11}$$

where $\Psi_c[x(t)]$ is the continuous-time energy operator, and $x(t)$ is a single component signal.

To discretize this *continuous-time energy operator*, replace t by nT (T is the sampling period), $x(t)$ by $x(nT)$ or simply $x[n]$, $\dot{x}(t)$ by its first backward difference, $y[n] = \frac{x[n]-x[n-1]}{T}$ and $\ddot{x}(t)$ by $\frac{y[n]-y[n-1]}{T}$. Then $\Psi_d(x[n])$, the *discrete-time energy operator* (the counterpart of the *continuous-time energy operator* $\Psi_c[x(t)]$) for discrete-time signal $x[n]$ is defined as,

$$\Psi_d(x[n]) = x^2[n] - x[n-1]x[n+1] \tag{4.12}$$

To simplify the notations, we henceforth drop the subscripts from the continuous and discrete energy operator symbols and use Ψ for both.

4.4 Energies of Well-Known Signals

Teager energy of some well known signals like sinusoidal, exponential, AM, FM and AM–FM signals are obtained in this section.

4.4.1 Sinusoidal Signal

Teager energy of the sinusoidal signal, $x(t) = Acos(\Omega t)$ is calculated by putting $x(t) = Acos(\Omega t)$ in Eq. 4.11,

$$\Psi[x(t)] = (-A\Omega sin(\Omega t))^2 - A cos(\Omega t)(-\Omega^2 A cos(\Omega t))$$
$$= A^2\Omega^2(sin^2(\Omega t) + cos^2(\Omega t))$$
$$= A^2\Omega^2 \tag{4.13}$$

which is the product of squared amplitude and frequency.

4.4.2 Exponential Signal

The exponentially decaying signal is modeled as $x(t) = e^{-\alpha t}$. Its Teager energy obtained using Eq. 4.11 is,

$$\Psi[e^{-\alpha t}] = (\alpha e^{-\alpha t})(\alpha e^{-\alpha t}) - e^{-\alpha t}(\alpha^2 e^{-\alpha t}) = 0 \tag{4.14}$$

Thus, the exponentially decaying signal has zero Teager energy.

4.4.3 AM Signal

AM signal modeling is described in Sect. 2.7.1. The Teager energy of an AM signal is,

$$\Psi[s_{AM}(t)] = \dot{a}^2(t)cos^2(\Omega_c t) + a^2(t)\Omega_c^2 - a(t)cos^2(\Omega_c t)\ddot{a}(t)$$
$$= a^2(t)\Omega_c^2 + cos^2(\Omega_c t)\Psi_c(a(t)) \tag{4.15}$$

We can see that the Teager energy of an AM signal is composed by a term similar to the energy of a sinusoidal signal, and an oscillation scaled by the Teager Energy of the amplitude signal. Figure 4.1 shows a sample AM signal (a) and its Teager energy (b). Notice the similarity between the envelope of the AM signal and the output of the Teager energy Operator.

4.4.4 FM Signal

The Teager Energy of an FM signal can be found to be

$$\Psi[s_{FM}(t)] = A^2\left(\dot{\phi}^2(t) + \ddot{\phi}\frac{sin(2\phi(t))}{2}\right) \tag{4.16}$$

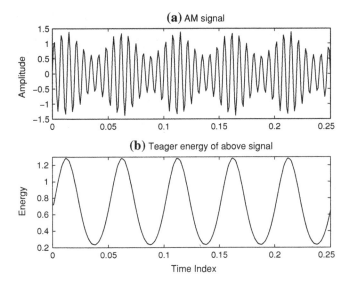

Fig. 4.1 **a** AM signal and **b** it's Teager energy

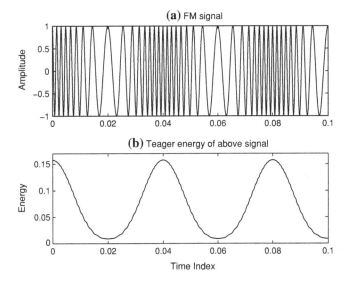

Fig. 4.2 **a** FM signal and **b** it's Teager energy

In Fig. 4.2, a sample FM signal and it's Teager energy is plotted. Here, the baseband signal is a pure sinusoidal signal. Notice that the output of the Teager Energy Operator in this case is a sinusoidal signal with the same frequency as the baseband signal.

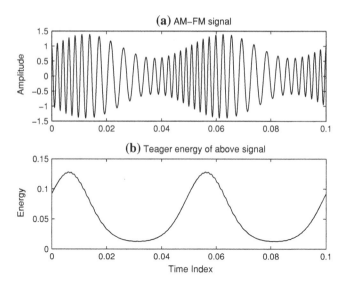

Fig. 4.3 **a** AM–FM signal and **b** it's Teager energy

4.4.5 AM–FM Signal

The Teager Energy of an AM–FM signal is

$$\Psi[s_{AM-FM}(t)] = [a(t)\phi(t)]^2 + \underbrace{\frac{1}{2}a^2(t)\ddot{\phi}(t)sin(2\phi(t))}_{FM} + \underbrace{cos^2(\phi(t))\Psi_c[a(t)]}_{AM}$$

(4.17)

In Fig. 4.3, an AM–FM signal and it's Teager energy is plotted. In this case, the operator's tracking capabilities are not apparent, although we can spot some correlation between the peaks of the Teager energy and the zero-crossings of the AM–FM signal.

4.5 Generalization of Teager Energy Operator

The discrete version of the Teager Operator can be generalized further. Wei Lin et al. [3] generalized the operator by replacing '1' with a constant M in the Eq. 4.12. Then it can be written as,

$$\Psi(x[n]) = x^2[n] - x[n - M]x[n + M]$$

(4.18)

The constant M is called the lag parameter. Choi-JH et al. used this generalization to resolve two closely spaced tones for action potential detection in neural research

[4]. The operator is generalized further by Moore et al. [5]. They have shown that the Teager energy operator can be approximated by a discrete high-pass Volterra system. The discrete version of a Volterra filter can be defined as,

$$y[n] = \sum_{i=0}^{N-1} x[n-i]h_1[i] \qquad (4.19)$$

where $h_1[i]$ is the impulse response (or Volterra kernel), and N is the length of h_1. This is nothing but the conventional convolution used to describe FIR filters. The two-dimensional discrete Volterra filter is then defined as,

$$y[n] = \sum_{j=0}^{N-1}\sum_{i=0}^{N-1} x[n-i]x[n-j]h_2[i,j] \qquad (4.20)$$

where $h_2[i,j]$ is the two-dimensional Volterra kernel.

In order to obtain the similarity between the two-dimensional Volterra system and the Teager energy operator, we can rewrite Eq. 4.12 as,

$$y[n] = x[n]x[n] - x[n-1]x[n+1] \qquad (4.21)$$

This is almost a two-dimensional Volterra system with the kernel h_2 defined as,

$$\begin{bmatrix} 0 & 0 & -1 \\ 0 & 1 & 0 \\ 0 & 0 & 0 \end{bmatrix}$$

This kernel is not symmetric, but we can make it symmetric by balancing the coefficients as,

$$\begin{bmatrix} 0 & 0 & -\frac{1}{2} \\ 0 & 1 & 0 \\ -\frac{1}{2} & 0 & 0 \end{bmatrix}$$

Notice that using the Volterra definition above, $h_2[i,j]$ is only defined for $i \geq 0$ and $j \geq 0$. This means that the output of the system is one sample delayed, because we require the Volterra system to be causal. Thus Teager energy operator can be expressed as a discrete quadratic Volterra-filter, if the kernel h_2 is chosen properly.

Mitra et al. have shown that the Teager energy operator can be approximated by a high-pass filter weighted by a local mean [6]. It is shown that,

$$y[n] \approx \mu\{-x[n-1] + 2x[n] - x[n+1]\} \qquad (4.22)$$

where μ is the local mean, $\mu = \{x[n-1] + x[n] + x[n+1]\}/3$. This mean value can be adjusted by changing the exponents in the original Teager energy operator as,

$$y[n] = x[n]^{\frac{2}{m}} - \{x[n-1]x[n+1]\}^{\frac{1}{m}} \qquad (4.23)$$

For $m = 1$, the extension reduces to the conventional Teager energy operator. The approximation is valid if the local mean is much larger than the variance of the signal. This means that the approximation is periodically incorrect for most audio signals, which usually are zero-mean signals [7]. By combining the two methods above, the discrete version can be written as,

$$y[n] = x[n]^{\frac{2}{m}} - \{x[n-M]x[n+M]\}^{\frac{1}{m}}. \qquad (4.24)$$

4.6 Energy Separation

We have seen that the Teager energy operator, when applied to an AM–FM sinewave, yields the squared product of an AM and FM components (i.e., amplitude envelope and instantaneous frequency). We now describe an approach to separate the time-varying amplitude envelope $a(t)$ and instantaneous frequency $\Omega_i(t)$ of an arbitrary AM–FM signal, based on the energy operator. First, we discuss a continuous-time solution for exact estimation of the constant amplitude and frequency of a sinewave and then show that the same equations approximately apply to an AM–FM signal with time varying amplitude and frequency. These algorithms are called as *energy separation algorithms* because an oscillator's energy depends on the product of amplitude and frequency [8]. Then, we discuss the energy separation algorithms for discrete-time signals.

4.6.1 Energy Separation for Continuous-Time Signals

Consider a sinusoidal signal $x(t) = A cos(\Omega_c t + \theta)$ with constant amplitude A and frequency Ω_c. It's Teager energy is calculated in Eq. 4.13 as, $\Psi[x(t)] = A^2\Omega_c^2$. Now consider it's derivative, $\dot{x}(t) = -A\Omega_c sin(\Omega_c t + \theta)$. It's Teager energy is,

$$\begin{aligned}
\Psi[\dot{x}(t)] &= \Psi[-A\Omega_c sin(\Omega_c t + \theta)] \\
&= A^2\Omega_c^4 cos^2(\Omega_c t + \theta) - [-A\Omega_c sin(\Omega_c t)][A\Omega_c^3 sin(\Omega_c t)] \\
&= A^2\Omega_c^4 \qquad (4.25)
\end{aligned}$$

We can combine above two results to obtain the constant frequency and amplitude as,

$$\Omega_c = \sqrt{\frac{\Psi[\dot{x}(t)]}{\Psi[x(t)]}} \qquad (4.26)$$

$$A = \frac{\Psi[x(t)]}{\sqrt{\Psi[\dot{x}(t)]}} \qquad (4.27)$$

Now consider the more general AM–FM signal of the form $s_{AM-FM}(t) = a(t)$ $cos[\phi(t)]$ as seen in Sect. 2.7.3. Specifically, it has been shown that Teager energy of an AM–FM signal can approximately estimate the squared product of the amplitude $a(t)$ and the instantaneous frequency $\Omega_i(t)$ signals. i.e.,

$$\Psi\left[a(t)cos\left(\int_0^t \Omega_i(\tau)d\tau + \theta\right)\right] \approx [a(t)\Omega_i(t)]^2 \qquad (4.28)$$

assuming that the parameters $a(t)$ and $\Omega_i(t)$ do not vary very fast (time rate of change of value) or too greatly (range of value) in time compared to the carrier frequency Ω_c. The derivative of the AM–FM signal $s_{AM-FM}(t) = a(t)cos[\phi(t)]$ is obtained as,

$$\dot{s}_{AM-FM}(t) = \dot{a}(t)cos[\phi(t)] - a(t)\Omega_i(t)sin[\phi(t)] \qquad (4.29)$$

With the assumptions specified in [8], we can obtain it's Teager energy as,

$$\Psi[\dot{s}_{AM-FM}(t)] \approx \Psi[a(t)\Omega_i(t)sin\phi(t)] \approx a^2(t)\Omega_i^4(t) \qquad (4.30)$$

By combining above two equations, we can obtain,

$$\Omega_i(t) \approx \sqrt{\frac{\Psi[\dot{x}(t)]}{\Psi[x(t)]}} \qquad (4.31)$$

$$a(t) \approx \frac{\Psi[x(t)]}{\sqrt{\Psi[\dot{x}(t)]}} \qquad (4.32)$$

4.6.2 Energy Separation for Discrete-Time Signals

The discrete-time version of all obove results can be obtained by using a discrete-time Teager energy operator, $\Psi(x[n]) = x^2[n] - x[n-1]x[n+1]$. If $x[n]$ is a sampled version of a continuous-time signal and replacing derivatives $\dot{x}$ with 2 sample backward (or forward) differences $(x[n] - x[n-1])/T$, where T is the sampling period, then the continuous-time energy operator reduces to the discrete version as [8],

$$\Psi(x[n]) = (x^2[n] - x[n+1]x[n-1])/T^2 \qquad (4.33)$$

Let us consider a constant amplitude/frequency discrete time cosine as, $x[n] = Acos(\omega_c n + \theta)$ where $\omega_c = \Omega_c T$ and $\Omega_c < \pi/T$. Then,

$$\Psi(x[n]) = A^2 sin^2(\omega_c)/T^2 \tag{4.34}$$

which can be rewritten as,

$$\Psi(x[n]) = A^2\Omega_c^2 \left(\frac{sin\omega_c}{\omega_c}\right)^2 \tag{4.35}$$

It shows that, in comparison with the continuous time case, the discrete-time case has additional attenuating term $(sin\omega_c\backslash\omega_c)^2$. Thus the two results are similar. Since ω_c carries the information about T, we can assume $T = 1$. This assumption is also correct for signals that are inherently defined only for discrete time [8].

Now, consider the first derivative obtained using the backward difference as,

$$\begin{aligned} y[n] &= x[n] - x[n-1] \\ &= A(cos[\omega_c n + \theta] - cos[\omega_c[n-1]] + \theta) \end{aligned} \tag{4.36}$$

using the following trigonometric identity,

$$cos(\alpha) - cos(\beta) = 2sin\left(\frac{\alpha+\beta}{2}\right) sin\left(\frac{\beta-\alpha}{2}\right) \tag{4.37}$$

we can write,

$$y[n] = -2Asin[\omega_c/2]sin[\omega_c n + \theta - \omega_c/2] \tag{4.38}$$

Using the Teager energy operator on this function yields

$$\Psi(y[n]) = 4A^2 sin^2(\omega_c/2)sin^2(\omega_c) \tag{4.39}$$

From Eqs. 4.35 and 4.39,

$$\frac{\Psi(y[n])}{2\Psi(x[n])} = 2sin^2(\omega_c/2) = 1 - cos(\omega_c). \tag{4.40}$$

From this, we can derive the equations for absolute amplitude and constant frequency as,

$$\omega_c = arccos\left(1 - \frac{\Psi(x[n] - x[n-1])}{2\Psi(x[n])}\right) \tag{4.41}$$

$$|A| = \sqrt{\frac{\Psi(x[n])}{sin^2[\omega_c]}}$$

$$= \sqrt{\frac{\Psi(x[n])}{1 - cos^2[\omega_c]}}$$

$$= \sqrt{\frac{\Psi(x[n])}{1 - \left(1 - \frac{\Psi(x[n]-x[n-1])}{2\Psi(x[n])}\right)^2}} \qquad (4.42)$$

We can also find the frequency and amplitude by using the above equations and replacing $y[n]$ with the forward difference $x[n+1] - x[n]$.

Now, specifically, consider an AM–FM discrete-time sinewave of the form,

$$x[n] = a[n]cos(\phi[n])$$

$$= a[n]cos\left(\omega_c n + \omega_m \int_0^n q[m]dm + \theta\right) \qquad (4.43)$$

with instantaneous frequency,

$$\omega_i[n] = \frac{d}{dn}\phi[n] = \omega_c + \omega_m q[n] \qquad (4.44)$$

where $|q[n]| \leq 1$, $\omega_m \in [0, \omega_c]$ is the frequency deviation and θ is a constant phase offset.

A number of different discrete-time energy separation algorithms have been derived using different approximating derivatives and different assumptions on the form of the AM and FM functions [8]. For AM–FM signal, we can obtain the instantaneous frequency and envelope estimated as

$$\omega_i[n] \approx arccos\left(1 - \frac{\Psi(x[n] - x[n-1])}{2\Psi(x[n])}\right) \qquad (4.45)$$

$$|a[n]| = \sqrt{\frac{\Psi(x[n])}{1 - \left(1 - \frac{\Psi(x[n]-x[n-1])}{2\Psi(x[n])}\right)^2}} \qquad (4.46)$$

This algorithm is called as DESA-1a (Discrete energy separation algorithm-1a), where "1" implies the approximation of derivatives with a single sample difference and "a" refers to use of asymmetric difference.

If we consider the symmetric difference of the form

$$y[n] = \frac{x[n+1] - x[n-1]}{2}, \qquad (4.47)$$

we can obtain following formulas for estimating the time-varying frequency and amplitude envelope as [9],

$$\omega_i[n] \approx arcsin\left(\sqrt{\frac{\Psi(x[n+1]) - x[n-1]}{4\Psi(x[n])}}\right) \tag{4.48}$$

$$|a[n]| \approx \frac{2\Psi(x[n])}{\sqrt{\Psi(x[n+1]) - x[n-1]}} \tag{4.49}$$

This is called as DESA-2 algorithm, "2" implies the approximation of the first-order derivatives by difference between samples whose time indices differ by 2.

An alternative algorithm results if we replace the derivatives with backward and forward difference,

$$y[n] = x[n] - x[n-1]; \quad z[n] = x[n+1] - x[n] = y[n+1]$$

and working as for DESA-2, we can obtain following formulas for estimating the time-varying frequency and amplitude envelope as [9],

$$\omega_i[n] \approx arccos\left(1 - \frac{\Psi(y[n]) + \Psi(y[n+1])}{4\Psi x([n])}\right) \tag{4.50}$$

$$|a[n]| \approx \sqrt{\frac{\Psi(x[n])}{1 - \left(1 - \frac{\Psi(y[n]) + \Psi(y[n+1])}{4\Psi x([n])}\right)^2}} \tag{4.51}$$

This algorithm is called as DESA-1 algorithm, where "1" implies the approximation of derivatives with a single sample difference. In [8], the (mean absolute and rms) errors of the two DESAs in estimating the amplitude and the frequency of synthetic AM–FM signals are compared by Maragos et al. On an average (for AM–FM amounts of 5–50%), both DESAs yielded very small errors of the order of 1% or less. In the presence of added white Gaussian noise at a 30 dB SNR, the DESA-2 algorithm yields errors of the order of 10% or less [10].

By assuming that the time-varying amplitude and frequency modulating signals can be modeled by a sum of slowly varying sinusoids within a pitch period, i.e., the AM and FM do not vary too fast in time or too greatly compared to the carrier ω_c, it follows that the energy separation algorithm can be applied to estimate AM and FM variations within a glottal cycle [10]. However, the AM–FM model of a single resonance does not explicitly take into considerations that, actual speech vowels are quasi-periodic and usually consist of multiple resonances. Various approaches can be found in [11] for jointly estimating the multiple AM–FM sinewave components of speech signals.

4.7 Teager Energy Operator in Noise

Consider a signal $s[n]$ degraded by zero-mean white Gaussian noise $v[n]$ with variance σ^2 as,

$$x[n] = s[n] + v[n] \tag{4.52}$$

The Teager energy of the noisy signal $x[n]$ is given by,

$$\Psi(x[n]) = \Psi(s[n]) + \Psi(v[n]) + 2\tilde{\Psi}(s[n], v[n]) \tag{4.53}$$

where, $\Psi(s[n])$ and $\Psi(v[n])$ are the Teager energies of the speech signal and the additive noise, respectively. $\tilde{\Psi}(s[n], v[n])$ is the cross-Ψ energy of $s[n]$ and $v[n]$ such that

$$\tilde{\Psi}(s[n], v[n]) = s[n]v[n] - (1/2)s[n-1]v[n+1] - (1/2)s[n+1]v[n-1] \tag{4.54}$$

Since $s[n]$ and $v[n]$ are independent, the expected value of their cross-Ψ energy is zero, therefore we can write,

$$\begin{aligned} E\{\Psi(x[n])\} &= E\{\Psi(s[n])\} + E\{\Psi(v[n])\} \\ &= E\{\Psi(s[n])\} + \sigma^2 \end{aligned} \tag{4.55}$$

It shows that the Teager energy estimate is biased by the variance of the noise.

4.7.1 Noise Suppression Using Teager Energy Operator

Let $s[n]$ be a discrete-time wide-sense stationary random signal. The expected value of it's Teager energy is,

$$\begin{aligned} E\{\Psi(s[n])\} &= E\{s^2[n])\} - E\{s[n+1]s[n-1]\} \\ &= R_s(0) - R_s(2) \end{aligned} \tag{4.56}$$

where $R_s(0)$ is the autocorrelation function of $s[n]$. It is shown in [12] that, the TEO has filtering capability.

Now consider an example of a car engine noise, $v[n]$. Figure 4.4 shows the PSD of car engine noise and it's Teager energy. It shows that, the car engine noise is mostly low pass in nature. The relation between the first three autocorrelation lags as obtained in [12] is $R_v(1) = 0.9999R_v(0)$, $R_v(2) = 0.9997R_v(1)$. Since $R_v(0) \approx R_v(1) \approx R_v(2)$, we have, $E\{\Psi(x[n])\} \approx 0$. Therefore the spectrum of $E\{\Psi(x[n])\}$ is almost negligible compared to the spectrum of $v[n]$, as shown in Fig. 4.4. Figure 4.5 shows the similar plots for babble noise.

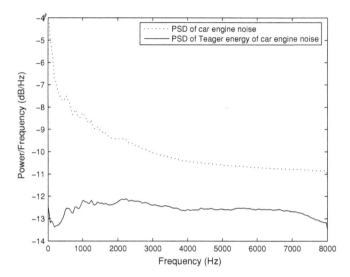

Fig. 4.4 The PSD of car engine noise and it's Teager energy

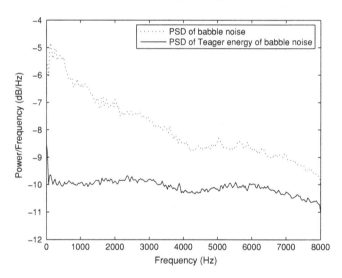

Fig. 4.5 The PSD of babble noise and it's Teager energy

Now carry the similar analysis for a speech signal. It shows that, the first three autocorrelation lags are not as close to each other. For example, for /s/, the autocorrelation lags are, $R_s(1) = 0.1541 R_s(0)$, $R_s(2) = 0.1805 R_s(1)$. For the signal /f/, the autocorrelation lags are, $R_s(1) = 0.4936 R_s(0)$, $R_s(2) = 0.2520 R_s(1)$. For another signal /aa/, the values are, $R_s(1) = 0.9294 R_s(0)$, $R_s(2) = 0.7377 R_s(1)$. Similar values for autocorrelation function for lag 0, 1 and 2 are obtained in [12]. Therefore, if we obtain a signal as a combination of speech and noise signals, then it is quite expected that, TEO should suppress noise.

4.8 Summary

In this chapter we have presented the nonlinear energy-tracking operator, i.e., the Teager energy operator for estimating the time-varying amplitude and instantaneous frequency of an AM-FM signal as well as the noise suppression capability of TEO is discussed with examples of car engine noise and babble noise. The only constraint for using TEO as an energy separator for AM-FM signal is that, the amplitude and frequency signals do not vary too fast with time compared with the carrier signal.

References

1. Kaiser JF (1993) Some useful properties of Teagers energy operator. In: Proceedings of IEEE international conference on acoustics, speech, and signal processing, vol 3, pp 149–152
2. Kaiser JF (1990) On a simple algorithm to calculate the energy of a signal. In: Proceedings of IEEE international conference on acoustics, speech, and signal processing, Albuquerque, NM, pp 381–384
3. Lin W, Hamilton C, Chitrapu P (1995) A generalization to the Teager-Kaiser energy function and application to resolving two closely-spaced tones. In: Proceedings of IEEE international conference on acoustics, speech, and signal processing (ICASSP'95), Detroit, MI, pp 1637–1640
4. Choi JH, Kim T (2002) Neural action potential detector using multi-resolution teo. Electron Lett 38:541–543
5. Moore M, Mitra S, Bernstein R (1997) A generalization of the Teager algorithm. In: Proceedings of IEEE workshop on nonlinear signal processing. Ann Arbor, MI
6. Mitra SK, Li H, Lin IS, Yu TH (1991) A new class of nonlinear filters for image enhancement. In: Proceedings of IEEE international conference acoustics, speech, and signal processing (ICASSP'91), Toronto, pp 2525–2528
7. Kvedalen E (2003) Signal processing using the Teager energy operator and other nonlinear operators. Candy Scientific Thesis, University of Oslo, Norway
8. Maragos P, Kaiser JF, Quatieri TF (1993) Energy separation in signal modulations with application to speech analysis. IEEE Trans Signal Process 41(10):3024–3051
9. Maragos P, Kaiser JF, Quatieri TF (1992) On separating amplitude from frequency modulations using energy operators. In: Proceedings of IEEE international conference on acoustics, speech, and signal processing (ICASSP'92), San Francisco, CA, pp 1–4
10. Quatieri TF (2004) Discrete-time speech signal processing, principles and practice. Pearson Education, Upper Saddle River
11. Torres WP, Quatieri TF (1999) Estimation of modulation based on am-fm transduction: two sinusoid case. IEEE Trans Signal Process 47(11):3084–3097
12. Jabloun F, Cetin AE, Erzin E (1999) Teager energy based feature parameters for speech recognition in car noise. IEEE Signal Process Lett 6(10):159–261

Chapter 5
AM-FM: Modulation and Demodulation Techniques

5.1 Introduction

Analysis of speech signals is usually carried out using STFT. The most successful features currently being used in both speech recognition and speaker recognition systems are cepstral features. The cepstral features in one way or another are based on the source-filter model of speech production [1, 2]. However, it is well known that a significant part of the acoustic information cannot be modeled by the linear source-filter model [3]. The source-filter model assumes that the sound source for the voiced speech is localized in the larynx and the vocal tract acts as a convolution filter for the emitted sound. Examples of phenomena not well-captured by the source-filter model include unstable airflow, turbulence and nonlinearities arising from oscillators with time-varying masses [3–5].

The cepstral features are computed from the magnitude spectrum of each frame of data, while leaving out the phase spectra. The approach of considering only the magnitude and discarding the phase is accepted so far, because the conventional analysis-synthesis methods believe that, the phase information is not important. However, Patterson [6] and Paliwal et al. [7–9] have shown that phase information is also important for audio perception. The approach to overcome the problem of neglecting the phase spectra is not to use the Fourier transform directly and instead use an AM-FM model introduced by Loughlin et al. [10] to represent a time varying speech signal as a sum of amplitude modulated (AM) and frequency modulated (FM) signals.

The AM-FM modeling technique has been applied to speech signal analysis with varying degrees of success, in areas such as formant tracking [11], speech synthesis [12], speech recognition [13, 14] and speaker identification [15]. It was first proposed by Potamianos et al. [11] in the context of formant tracking. The authors demonstrated that the AM-FM approach can overcome some of the limitations of the classic source-filter model and greatly help to the difficult task of speech formant tracking. Marco Grimaldi et al. [16] extended the same work to the problem of speaker identification. They indicated that the characterization of the different instantaneous frequencies

R. S. Holambe and M. S. Deshpande, *Advances in Non-Linear Modeling for Speech Processing*, SpringerBriefs in Speech Technology, DOI: 10.1007/978-1-4614-1505-3_5, © The Author(s) 2012

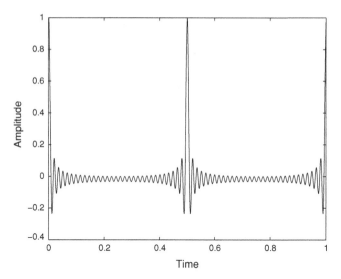

Fig. 5.1 Sum of 32 cosines

within the speech signal play a significant role in capturing the identity of a speaker. Jankowski et al. [15] adopted the AM-FM model to characterize some fine structures of the human voice for the purpose of speaker identification and demonstrated that the formant AM-FM parameters substantially improve identification rates on female speakers [15].

5.2 Importance of Phase

The cepstral features (MFCC and LPCC) are computed from the magnitude spectrum of each frame of data and the phase spectra is neglected. Neglecting the phase spectra, results in a loss of all information in the spectral content changes that occurs within the duration of a single frame. The delta cepstral features, which are designed to capture spectral changes between frames, also fail to capture such changes.

To show the importance of phase, Eric Lindemann et al. [17], compared two synthesized signals. One signal is the sum of harmonically related cosines. This signal is shown in Fig. 5.1, and is equivalent to a bandlimited pulse train, which might be used to synthesize the voiced excitation of a linear predictive speech synthesizer.

The other signal is a sum of harmonically related cosines with random initial phase as shown in Fig. 5.2.

Both of these signals are perfectly periodic and have identically constant magnitude spectra. Yet they sound different. The random-phase signal sounds more 'active' in the high frequencies with a less pronounced fundamental. The authors shown that, the phase relationships between high frequency sinusoids in a critical band affect the

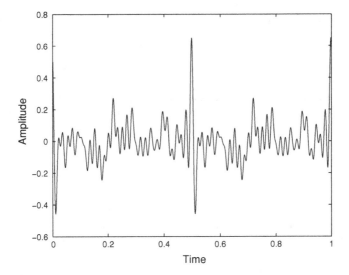

Fig. 5.2 Sum of 32 cosines with random phase

signal envelope and, as a result, the firing rates of inner hair cells associated with the critical band. Therefore, sounds with identical magnitude spectra can result in different firing patterns and causes the difference in perception. Fan-Gang Zeng et al. conducted listening tests using stimuli with different modulations in normal-hearing and cochlear-implant subjects [18]. They found that although AM from a limited number of spectral bands may be sufficient for speech recognition in quiet, FM significantly enhances speech recognition in noise, as well as speaker recognition. Further the importance of phase is discussed by Paliwal et al. in [8, 9].

5.3 AM-FM Model

AM-FM model is a technique used especially by electrical engineers in the context of frequency modulated signals, such as FM radio signals. It can be effectively used for modeling the speech production system. Vocal tract resonances can change rapidly both in frequency and amplitude even within a single pitch period. This may be due to rapidly varying and separated speech airflow in the vocal tract [4]. The effective air masses in vocal tract cavities and effective cross sectional areas of the airflow vary rapidly, causing modulations of air pressure and volume velocity. This leads to the actual speech signal, $s(t)$ composed of a sum of N resonances as,

$$s(t) = \sum_{i=1}^{N} R_i(t) \tag{5.1}$$

where $R(t)$ is a single speech resonance, which can be represented as an AM-FM signal,

$$R(t) = a(t)cos\left[2\pi(f_c t + \int_0^t q(\tau)d\tau) + \theta\right] \tag{5.2}$$

where f_c is the center value of the resonance (formant) frequency, $q(t)$ is the frequency modulating signal and $a(t)$ is the time varying amplitude. The individual resonances may be isolated by band-pass filtering the speech signal. The instantaneous resonance frequency signal is defined as,

$$f_i(t) = f_c + q(t). \tag{5.3}$$

The estimation of the amplitude envelope and instantaneous frequency components, i.e., the demodulation of each resonant signal, can be done with the energy separation algorithm (ESA), or utilizing the Hilbert transform demodulation (HTD) algorithm as described in the next section.

Above discussion shows that, the speech signal (obtained through a speech production system) is of the type of AM-FM signal. Further, from speech perception viewpoint, the hypothesis given by Saberi and Hafter [19] for the measurement of frequency modulation by the auditory system is that the cochlear filters, and perhaps higher level neurophysiological tuning curves, use transduction of frequency modulation (FM) to amplitude modulation (AM); the instantaneous frequency of the FM sweeps through the nonflat passband of the filter, thus inducing a change in the amplitude envelope of the filter output. Psychoacoustic experiments by Saberi and Hafter indicate that FM and AM may be transformed into a common neural code in the brain stem [19]. Therefore, it is quite expected that, if we obtain a feature set based on speech production as well as speech perception mechanism, it will be more robust.

5.3.1 Amplitude Modulation and Demodulation

The amplitude modulation process is shown in Fig. 5.3 which contains the baseband signal, carrier signal and the amplitude modulated signal. From this figure, we can clearly see that the baseband signal is the envelope of the carried wave in the AM signal. It is called as amplitude modulation because the amplitude of the carrier is modified.

The demodulation process of AM signals is basically to extract the envelope of the carrier wave in the AM signal. This can be obtained by using an envelope detector [20], which is shown in Fig. 5.4. On the positive half-cycle of the input signal, the diode, D is forward-biased, and the capacitor, C charges up rapidly to the input signal. When the input signal falls below its peak value, the capacitor discharges slowly through the resistor R_L until the next positive half-cycle. It results into the

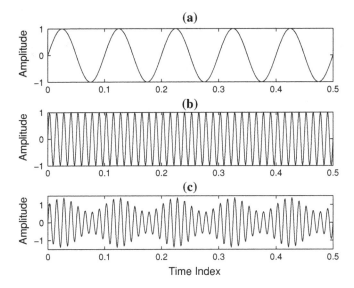

Fig. 5.3 Amplitude modulation process, **a** baseband signal, **b** carrier signal and **c** amplitude modulated signal

Fig. 5.4 Envelope detector
used to extract baseband signal
from the AM signal

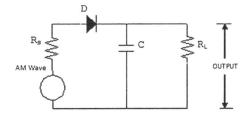

envelope of the carrier with ripples. The ripple is then removed by a low-pass filter, and we can obtain the baseband signal.

5.3.2 Frequency Modulation and Demodulation

Figure 5.5 shows the frequency modulation process. In this example we can clearly see that the frequency of the carrier signal varies according to the amplitude of the baseband signal. The demodulation process for FM signals is more complicated than for the AM signals. The demodulation can be done using several methods like slope circuit and envelope detector or phase-locked loop [20]. Let's consider the slope circuit and envelope detector method. Figure 5.6 shows a block schematic of this method which consists of a slope circuit and an envelope detector.

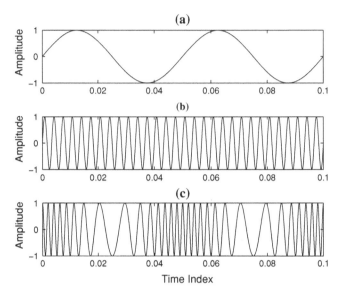

Fig. 5.5 Frequency modulation process, **a** baseband signal, **b** carrier signal and **c** frequency modulated signal

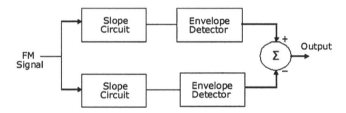

Fig. 5.6 Slope circuit and envelope detector used for FM demodulation

The slope circuit is essentially a band pass filter, which effectively converts the FM signal into an AM-FM signal, where the AM part of the signal is the modulating signal in the original FM signal. The envelope detector is then used to extract the amplitude, and this is how we can obtain the baseband signal, but it is biased by the FM part. Of course, the bias can be removed by creating a mirrored system which computes a signal with a bias which when combined with the first signal produces an unbiased signal.

5.4 Estimation Using the Teager Energy Operator

As shown in the previous section, the speech signal can be represented as an AM-FM signal. Estimating two important parameters i.e., amplitude envelope and instantaneous frequency of an AM-FM signal can be done by demodulating the AM-FM signal. In this section, we shall describe a method of how to estimate the parameters of AM and/or FM signals using the Teager energy operator. As shown in Sect. 4.6.1, for a single sinusoidal signal, $x(t) = A\cos(\Omega_c t + \theta)$, it's amplitude and frequency can be estimated using Teager energy operator as,

$$\Omega_c = \sqrt{\frac{\Psi[\dot{x}(t)]}{\Psi[x(t)]}}$$

$$A = \frac{\Psi[x(t)]}{\sqrt{\Psi[\dot{x}(t)]}}$$

These equations are exact for a sinusoidal signal, but there are small approximation errors for AM, FM, and AM-FM signals. However, as the errors are small, the above method is suitable for estimation of the envelope of AM signals and the instantaneous frequency of FM signals.

Consider an AM signal as described in Sect. 2.7.1 and it's Teager energy obtained in Sect. 4.4.3 as,

$$\Psi[s_{AM}(t)] = \underbrace{a^2(t)\Omega_c^2}_{D(t)} + \underbrace{cos^2(\Omega_c t)\Psi_c(a(t))}_{E(t)}$$

the term Ω_c is a constant in the expression above. If $E(t) \ll D(t)$, then the output is a constant multiplied by the envelope of the AM signal. We can look at $E(t)$ as an estimation error, and if this error is sufficiently small, the output from the Teager energy operator tracks the envelope of the AM signal nicely. This proof is rather long and non-trivial and details are covered in the article by Maragos et al. [4].

In case of an FM signal as defined in Sect. 2.7.2 and it's Teager energy is obtained as (Eq. 4.16),

$$\Psi[s_{FM}(t)] = \underbrace{A^2\dot{\phi}^2(t)}_{D(t)} + \underbrace{A^2\ddot{\phi}\frac{sin(2\phi(t))}{2}}_{E(t)}$$

If the term $E(t)$ is small enough, i.e., $E(t) \ll D(t)$, we can ignore it, and we then see that the operator tracks the instantaneous frequency squared, multiplied by a constant. Just as for AM signals, the proof is rather long and can be found in [4].

When AM and FM signals are combined, it can be seen from Eq. 4.17 that, the Teager energy for the AM-FM signal contains the term $[a(t)\phi(t)]^2$, and if we can show that the other terms are small compared to $[a(t)\phi(t)]^2$, then we can use the Teager energy operator to estimate amplitude and instantaneous frequency of AM-FM signals. The details of this are covered by Maragos et al. in [4].

5.5 Estimation Using the Hilbert Transform

In order to characterize single instantaneous frequency for a real-valued signal, an analytic signal is first constructed; it is a transformation of a real signal into the complex domain. More formally, given a real input signal $s(t)$, its analytic signal $s_a(t)$ can be computed as,

$$s_a(t) = s(t) + j\hat{s}(t) \tag{5.4}$$

where $\hat{s}(t)$ is the Hilbert transform of $s(t)$. The Hilbert transform does not involve a domain change, i.e., Hilbert transform of a signal $s(t)$ is another signal $\hat{s}(t)$ in the same domain. The frequency components of $\hat{s}(t)$ lags the frequency components of $s(t)$ by 90^0. Obviously, performing the Hilbert transform on a signal is equivalent to a phase shift in all its frequency components. The most important fact is, the amplitude of the frequency components of the signal and therefore the energy and power of the signal, do not change by performing the Hilbert transform. We can decompose the analytical signal $s_a(t)$ as follows:

$$s_a(t) = a_i(t)e^{j\phi(t)} \tag{5.5}$$

where

$$a_i(t) = \mid s_a(t) \mid \tag{5.6}$$

is called the *instantaneous amplitude* (or Hilbert envelope) of the signal, and

$$\phi(t) = \angle s_a(t) = \arctan\left[\hat{s}(t)/s(t)\right] \tag{5.7}$$

is the *instantaneous phase*. This phase is in time domain and quite different from Fourier transform phase because, it is derived from analytic signal concept. The *instantaneous frequency* $f_i(t)$ is computed from the unwrapped *instantaneous phase* $\phi_u(t)$ as follows:

$$f_i(t) = \frac{1}{2\pi}\frac{d\phi_u(t)}{dt} \tag{5.8}$$

where $\phi_u(t) = \phi(t) + L(t)$, is the the continuous, unwrapped *instantaneous phase* and $L(t)$ is a integer multiple of π valued function designed to ensure a continuous phase function. Equations 5.6 and 5.8 show how to obtain instantaneous amplitude and frequency using HTD.

The instantaneous frequency estimation is one of the effective methods to detect and track frequency changes of a mono-component signal. But, in the case of multi-component signals, the result becomes meaningless without breaking the signal down into its components [21]. As discussed in [21], the decomposition of a signal is not unique if its frequency components coincide at some points in the time-frequency plane. This is the case for speech, e.g., formants are well known to have points in the time-frequency plane where they appear to join or split. In this case, the decomposition is heuristic in nature and its optimal form will depend on the specific application. In order to estimate the center formant frequency we need to separate that formant from the speech signal using a proper filter. In the optimal case, the bandwidth of this filter does not overlap on the neighboring formants and the filter center frequency is set at the center frequency of the formant.

5.6 Multiband Filtering and Demodulation

To obtain a single resonance signal $R(t)$, from the speech signal $s(t)$ (ref Eq. 5.1), a filtering scheme can be used before demodulation, which is referred as multiband demodulation analysis (MDA). The MDA yields rich time-frequency information. The MDA consists of a multi-band filtering scheme and a demodulation algorithm. First, the speech signal is band-pass filtered using a filterbank, then each band-pass waveform is demodulated and its instantaneous amplitude and frequency are computed.

The following steps are adopted to demodulate the speech signal and to extract the features.

- The speech signal $s(t)$ is band-pass filtered and a set of waveforms $w_k(t)$ is obtained (k denotes the kth filter in the filterbank).
- For each band-pass waveform $w_k(t)$, its Hilbert transform $\hat{w}_k(t)$ is computed.
- The instantaneous amplitude, $a_{ik}(t)$ for each band-pass waveform is computed as,

$$a_{ik}(t) = \sqrt{w_k^2(t) + \hat{w}_k^2(t)}. \tag{5.9}$$

- The instantaneous frequency, $f_{ik}(t)$ for each band-pass waveform is computed as the first time derivative of the unwrapped *instantaneous phase* $\phi_k(t)$ as,

$$f_{ik}(t) = \frac{1}{2\pi} \cdot \frac{d\phi_k(t)}{dt} = \frac{1}{2\pi} \cdot \frac{d}{dt} \left[arctan(\hat{w}_k(t)/w_k(t)) \right]. \tag{5.10}$$

After obtaining the instantaneous amplitude and frequency signals by demodulating each resonant signal, a short-time analysis is performed.

5.7 Short-Time Estimates

To derive robust features, after demodulating the speech signal, i.e., extracting the instantaneous amplitude and frequency, the short-time estimates of the frequency i.e., the first and second moment can be obtained. Obtaining weighted and unweighted first and second moment is discussed in the following sections.

5.7.1 Short-Time Estimate: Frequency

Simple short-time estimate of the frequency F is the unweighted mean F_{iu} of the instantaneous frequency signal $f_i(t)$, i.e.,

$$F_{iu} = \frac{1}{\tau} \int_{t_0}^{t_0+\tau} f_i(t)dt, \tag{5.11}$$

where t_0 and τ are the start and duration of the analysis frame, respectively. Alternative estimate is the first weighted moment of $f_i(t)$ [11, 22]. Using the squared amplitude, $a_i^2(t)$ as the weight, the first weighted moment is,

$$F_{iw} = \frac{\int_{t_0}^{t_0+\tau}[f_i(t)a_i^2(t)]dt}{\int_{t_0}^{t_0+\tau}[a_i^2(t)]dt}. \tag{5.12}$$

The adoption of a mean amplitude weighted instantaneous frequency is motivated by the fact that it provides more accurate frequency estimate and is more robust for low energy and noisy frequency bands when compared with an unweighted frequency mean as shown by Potamianos et al. in [11] as well as Dimitriadis et al. in [13].

To understand the importance of weighted moment, let's consider the example given in [11]. Consider the sum, $x(t)$ of two sinusoids with constant frequencies $f_1 = 1.5\,\text{kHz}$ and $f_2 = 1.7\,\text{kHz}$, and time-varying amplitudes $a_1(t), a_2(t)$:

$$x(t) = a_1(t)cos[2\pi f_1 t] + a_2(t)cos[2\pi f_2 t], t \in [0, 0.1]s. \tag{5.13}$$

where $a_1(t) = 10t$ and $a_2(t) = 1 - 10t$, therefore for the first half of the time interval (0 to 50 ms), the second sinusoid f_2 is dominant, while for the second half (50 to 100 ms), f_1 dominates.

Figure 5.7 shows the amplitude envelope $| a(t) |$ of $x(t)$ and Fig. 5.8 shows the instantaneous frequency $f_i(t)$ of $x(t)$ computed via Hilbert transform demodulation (HTD). At envelope maxima, the instantaneous frequency is equal to the average (amplitude weighted) frequency of the two sinusoids $f = (a_1 f_1 + a_2 f_2)/(a_1 + a_2)$, while at envelope minima, f presents spikes of value $f = (a_1 f_1 - a_2 f_2)/(a_1 - a_2)$; i.e., the spikes point toward the frequency of the sinusoid with the larger amplitude.

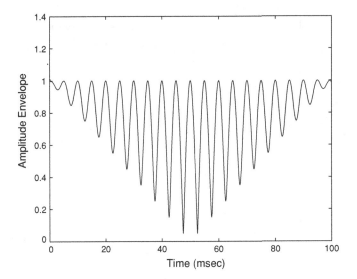

Fig. 5.7 Amplitude envelope of $x(t)$ obtained using HTD

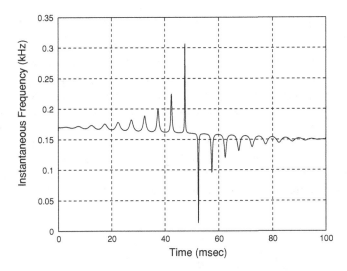

Fig. 5.8 Instantaneous frequency of $x(t)$ obtained using HTD

The short-time unweighted estimate, F_{iu} and weighted estimate, F_{iw} of the instantaneous frequency computed by the HTD are shown in Fig. 5.9. F_{iu} locks onto the sinusoid with the greater amplitude whereas F_{iw} provides a more 'natural' short-time estimate because the spikes of the instantaneous frequency correspond to amplitude minima, and get weighted less in the F_{iw} average. Actually, F_{iw} is the mean weighted frequency of the two sinusoids, with squared amplitude as the weight.

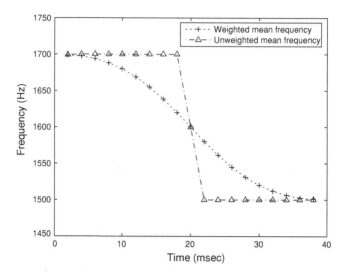

Fig. 5.9 Weighted and unweighted estimates of the instantaneous frequency of $x(t)$ obtained using HTD

These results can be generalized to the short-time frequency estimates of speech resonances by use of a sinusoidal speech model. A speech signal can be modeled as a sum of sinusoids with slowly time-varying amplitudes and frequencies [23]. In particular, a speech resonance can be modeled as a sum of a few sinusoids. The behavior of the F_{iu} and F_{iw} estimates for a speech formant can then be viewed as a generalization of the two sinusoids case analyzed above. For a speech resonance signal, F_{iu} has the tendency to lock on the frequency with the greatest amplitude in the formant band, while F_{iw} weights each frequency in the formant band with its squared amplitude. Thus the weighted frequency estimate F_{iw} provides more accurate formant frequencies and is more robust for low energy or noisy frequency bands [11].

5.7.2 Short-Time Estimate: Bandwidth

We have seen the short-time estimate for the frequency, now consider the short-time estimate for the bandwidth. The estimate of the bandwidth B is nothing but standard deviation B_{iu} of the instantaneous frequency signal $f_i(t)$, i.e.,

$$[B_{iu}]^2 = \frac{1}{\tau} \int_{t_0}^{t_0+\tau} [f_i(t) - F_{iu}]^2 dt, \qquad (5.14)$$

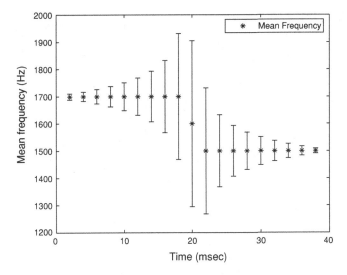

Fig. 5.10 The short-time unweighted estimates of the instantaneous frequency, F_{iu} and the bandwidth B_{iu} computed by the HTD. The bandwidths are shown as 'error bars' around their respective frequency estimates

where t_0 and τ are the start and duration of the analysis frame, respectively and F_{iu} is the unweighted mean instantaneous frequency as in Eq. 5.11. Alternative estimate is the second weighted moments of $f_i(t)$ as discussed by Potamianos et al. in [11, 22]. Using the squared amplitude, $a_i^2(t)$ as the weight, the second weighted moment is,

$$[B_{iw}]^2 = \frac{\int_{t_0}^{t_0+\tau}[(\dot{a}_i(t)/2\pi)^2 + (f_i(t) - F_{iw})^2 . a_i^2(t)]dt}{\int_{t_0}^{t_0+\tau}[a_i^2(t)]dt}, \qquad (5.15)$$

where the additional term $(\dot{a}_i(t)/2\pi)^2$ in $[B_{iw}]^2$ accounts for the amplitude modulation contribution of the bandwidth [11, 24] and F_{iw} is as in Eq. 5.12.

Now, consider the previous example, where we have obtained a signal $x(t)$ as a sum of two sinusoids with constant frequencies $f_1 = 1.5\,\text{kHz}$ and $f_2 = 1.7\,\text{kHz}$, and time-varying amplitudes $a_1(t), a_2(t)$. For the signal $x(t)$, the short-time unweighted estimates of the instantaneous frequency (F_{iu}) and the bandwidth (B_{iu}) are computed using HTD and shown in Fig. 5.10. Similarly, Fig. 5.11 shows the short-time weighted estimates of the instantaneous frequency (F_{iw}) and the bandwidth (B_{iw}) computed by the HTD. The bandwidths are shown as 'error bars' around their respective frequency estimates. Note that for $a_1 \approx a_2$ (i.e., when there is not a single prominent harmonic in the spectrum), B_{iu} takes unnaturally large values. The B_{iw} bandwidth estimate is more robust than the B_{iu} estimate. As B_{iw} equals the second spectral moment, it is by definition, the *rms bandwidth* of the signal.

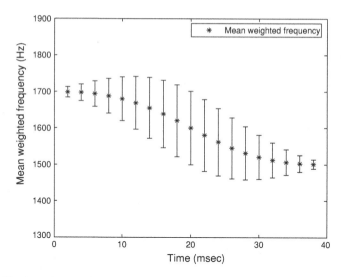

Fig. 5.11 The short-time weighted estimates of the instantaneous frequency, F_{iw} and the bandwidth B_{iw} computed by the HTD. The bandwidths are shown as 'error bars' around their respective frequency estimates

The estimates F_{iw} and B_{iw} can be computed in the frequency domain as the first and second spectral moments (computed via FFT). This results in significant computational savings since the filtering can be implemented by multiplication in the frequency domain and no demodulation is needed.

5.8 Summary

Psychoacoustic experiments by Saberi and Hafter [19] provide evidence that FM and AM sounds must be transformed into a common neural code before binaural convergence in the brain stem. They showed that, listeners can accurately determine if the phase of an FM signal presented to one ear is leading ar lagging the phase of an AM signal presented to the other ear, with peak performance occuring when the signals are 180^0 out of phase. Motivated by this hypothesis, an approach to AM-FM estimation is discussed in this chapter. Use of Teager energy operator and Hilbert transform to demodulated the signal to estimate the amplitude and instantaneous frequency is discussed. To obtain features from the demodulated signal, how short-time estimates can be useful is also presented in this chapter. It shows that the weighted estimates of the instantaneous frequency and bandwidth are more natural compared to the unweighted estimates.

References

1. Rabiner LR, Shafer RW (1989) Digital signal processing of speech signals. Prentice-Hall, Englewood Cliffs
2. Rao A, Kumaresan R (2000) On decomposing speech into modulated components. IEEE Trans Speech Audio Process 8:240–254
3. Dimitriadis D, Maragos P (2003) Robust energy demodulation based on continuous models with application to speech recognition. In: Proceedings of EUROSPEECH'03, Geneva, pp 2853–2856
4. Maragos P, Kaiser JF, Quatieri TF (1993) Energy separation in signal modulations with application to speech analysis. IEEE Trans Signal Process 41(10):3024–3051
5. Teager HM (1980) Some observations on oral air flow during phonation. IEEE Trans Speech Audio Process 28(5):599–601
6. Patterson RD (1987) A pulse ribbon model of monoaural phase perception. J Acoust Soc Am 82(5):1560–1586
7. Paliwal K, Arslan L (2003) Usefulness of phase spectrum in human speech perception. In: Proceeding of EUROSPEECH'03, Geneva, pp 2117–2120
8. Paliwal K, Alsteris L (2005) On the usefulness of stft phase spectrum in human listening tests. Speech Commun 45(2):153–170
9. Alsteris L, Paliwal K (2006) Further intelligibility results from human listening tests using the short-time phase spectrum. Speech Commun 48(6):727–736
10. Loughlin PJ, Tacer B (1996) On the amplitude and frequency modulation decomposition of signals. J Acoust Soc Am 100(3):1594–1601
11. Potamianos A, Maragos P (1996) Speech formant frequency and bandwidth tracking using multiband energy demodulation. J Acoust Soc Am 99(6):3795–3806
12. Li G, Qiu L, Ng LK (2000) Signal representation based on instantaneous amplitude models with application to speech synthesis. IEEE Trans Speech Audio Process 8(3):353–357
13. Dimitriadis V, Maragos P, Potamianos A (2005) Robust AM-FM features for speech recognition. IEEE Signal Process Lett 12(9):621–624
14. Potamianos A, Maragos P (2001) Time-frequency distributions for automatic speech recognition. IEEE Trans Speech Audio Process 9(3):196–200
15. Jankowski CR, Quatieri TF, Reynolds DA (1995) Measuring fine structure in speech: application to speaker identification. In: Proceedings of the IEEE international conference on acoustics, speech, and signal processing, pp 325–328
16. Grimaldi M, Cummins F (2008) Speaker identification using instantaneous frequencies. IEEE Trans Audio Speech Lang Process 16(6):1097–1111
17. Lindemann E, Kates JM (1999) Phase relationships and amplitude envelopes in auditory perception. In: Proceedings of the IEEE workshop on applications of signal processing to audio and acoustics, New Paltz, New York, pp 17–20
18. Zeng FG, Nie K, Stickney GS, Kong YY, Vongphoe M, Bhargave A, Wei C, Cao K (2005) Speech recognition with amplitude and frequency modulations. Proc Natl Acad Sci U S A 102(7):2293–2298
19. Saberi K, Hafter ER (1995) A common neural code for frequency and amplitude-modulated sounds. Nature 374:537–539
20. Haykin S (1994) Communication systems. Wiley, New York
21. Boashash B (1992) Estimating and interpreting the instantaneous frequency of a signal-part 1: fundamentals. Proc IEEE 80(4):519–538
22. Potamianos A, Maragos P (1995) Speech formant frequency and bandwidth tracking using multiband energy demodulation. In: Proceedings of the IEEE international conference on acoustics, speech, and signal processing (ICASSP'95), pp 784–787
23. McAulay RJ, Quatieri TF (1986) Speech analysis/synthesis based on a sinusoidal representation. IEEE Trans Acoustic Speech Signal Process 34:744–754
24. Cohen L, Lee C (1992) Instantaneous bandwidth. In: Boashash B (ed) Time frequency signal analysis-methods and applications, Longman Cheshire, London

Chapter 6
Application to Speaker Recognition

6.1 Introduction

Speaker recognition refers to a task of recognizing people by their voices. In speaker recognition, one is interested in extracting and characterizing the speaker-specific information embedded in speech signal. In a larger context, speaker recognition belongs to the field of biometrics, which refers to authenticating persons based on their physical and/or learned characteristics [1, 2]. There has long been a desire to be able to identify a person on the basis of his or her voice. For many years, judges, lawyers, detectives and law enforcement agencies have wanted to use forensic voice authentication to investigate a suspect or to confirm a judgment of guilt or innocence [3].

Being a non-invasive biometric means of identification, it is useful for applications ranging from physical access control for high security areas to control of access to personal computers and information systems as well as attendance monitoring in large enterprizes. It is also useful for applications like, transaction processing via telephone or computer (e.g. account access, funds transfer, bill payment, trading of financial instruments), credit card processing (address changes, balance transfers, loss prevention), teleconferencing etc.

For recognizing speakers, traditional feature extraction methods focus on the large spectral peaks caused by the movement of the vocal tract and emphasize on lower frequency bands. However, Wu et al. [4], Itakura et al. [5] and Mishra et al. [6] have shown that a rich amount of speaker individual information is contained in the high frequency bands. Conventional theories of speech production are based on the linear source-filter model. The features such as LPC, LPCC and MFCC, based on this theory assume that the airflow propagates in the vocal tract as a linear plane wave [7, 8]. However, Teager [9] suggested that the true source of sound production is actually the vortex-flow interactions, which are nonlinear and act as the primary source of sound excitation in the vocal tract during phonation. The conventional theories of speech production are going to neglect the nonlinear aspects of speech production, which are also speaker-specific.

R. S. Holambe and M. S. Deshpande, *Advances in Non-Linear Modeling for Speech Processing*, SpringerBriefs in Speech Technology, DOI: 10.1007/978-1-4614-1505-3_6, © The Author(s) 2012

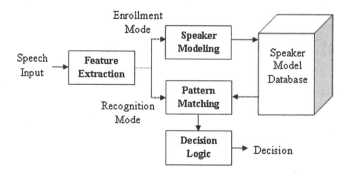

Fig. 6.1 Automatic speaker recognition system architecture

In this chapter, first, the importance of high frequency components in discriminating people is investigated and this information is fused with the commonly used Mel scale to derive a new filterbank. Then, the Teager energy operator (TEO) is used to extract the contribution of the speech fine structure resulting from the nonlinear aspects of speech production system. The proposed Teager energy operator based cepstral coefficient (TEOCC) features show significant improvement in the identification accuracy under mismatched training and testing environments. Then an AM–FM speaker model is used, which shows further improvement in the speaker identification rate in case of additive noise.

6.2 Speaker Recognition System

A speaker recognition system can not identify a speaker unless the system is trained with speech produced by that speaker, just as a human can not identify a person unless he/she already knows that person from previous encounters. Therefore a speaker recognition system has two operational modes: *enrollment* and *recognition*. In the enrollment mode, the user provides his/her voice sample to the system to train the model. In the recognition mode, the user provides another voice sample, which the system compares with the previously stored models and takes the decision.

Figure 6.1 shows the architecture of an automatic speaker recognition system. Regardless of the type of the task (*identification* or *verification*), both the enrollment and the recognition modes include *feature extraction* block, sometimes called the *front-end* of the system. The feature extractor converts the digital speech signal into a sequence of numerical descriptors, called *feature vectors*. The features provide a more stable, robust and compact representation than the raw input signal. Feature extraction can be considered as a data reduction process that attempts to capture the essential characteristics of the speaker. In the enrollment mode, a speaker model is created from the feature vectors and stored in the database.

In the recognition mode, features are extracted from the unknown speaker's voice sample. Pattern matching refers to an algorithm, or several algorithms, that compute a match score between the unknown speaker's feature vectors and the models stored in the database. The output of the pattern matching module is a similarity score. The last phase in the recognition chain is decision making. The decision module takes the match scores as its input and makes the final decision of the speaker identity, possibly with a confidence value [10, 11]. The type of the decision depends on the task i.e., verification or identification task. For the verification task, the binary decision is either acceptance or rejection of the speaker. The identification task can be of the type of open-set or closed-set identification. In closed-set identification, the decision is the ID number of the most similar speaker to the unknown speaker, whereas in the open-set identification, there is an additional decision that the speaker is none of the registered speakers.

The objective of modeling technique is to generate speaker models using speaker-specific feature vectors. Such models have enhanced speaker-specific information at reduced data rate. This is achieved by exploiting the working principles of the modeling techniques. There are two types of models: stochastic models and template models. In stochastic models, the pattern matching is probabilistic and results in a measure of likelihood of the observation for the given model. For template models, the pattern matching is deterministic. The observation is assumed to be an imperfect replica of the template and the alignment of the observed frames to template frame is carried out to minimize the distance. Further, the template method can be dependent or independent of time. Different modeling techniques used are Template matching [12–20], Vector Quantization (VQ) [21–27], Artificial Neural Network (ANN) [28–34], Gaussian Mixture Model (GMM) [35–37], Hidden Markov Model (HMM) [38–41], Support Vector Machines (SVM) [42–45] etc.

6.3 Preprocessing of Speech Signal

The digital speech signal is first preprocessed before extracting the features. The preprocessing involves pre-emphasis, framing and windowing operations.

6.3.1 Pre-Emphasis

Pre-emphasis refers to filtering, which emphasizes the higher frequencies. Its purpose is to balance the spectrum of voiced sounds that have a steep roll-off in the high frequency region. For voiced sounds, the glottal source has approximately $-12\,$dB/octave slope [46]. However, when the acoustic energy radiates from the lips, this causes a roughly $+6\,$dB/octave boost to the spectrum. As a net result, a speech signal when recorded with a microphone from a distance has approximately a $-6\,$dB/octave slope downward compared to the true spectrum (of the vocal tract).

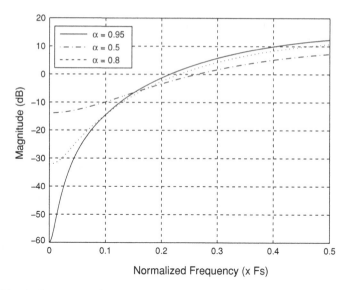

Fig. 6.2 Magnitude response of a pre-emphasis filter for different values of α

To eliminate this effect and to prevent lower frequency components from dominating the signal, pre-emphasis should be performed before feature extraction. The most commonly used pre-emphasis filter (first order FIR filter) is given by the following transfer function:

$$H(z) = 1 - \alpha z^{-1}, \qquad 0.9 \leq \alpha \leq 1.0 \tag{6.1}$$

where α controls the slope of the filter. The impulse response of the filter is $h[n] = \{1, -\alpha\}$ and the filter is simply implemented as a first order differentiator,

$$y[n] = s[n] - \alpha s[n-1]. \tag{6.2}$$

where $s[n]$ is the input signal and $y[n]$ is the output signal. The magnitude responses in dB scale for different values of α are shown in Fig. 6.2. An example of pre-emphasized speech frame in time and frequency domains is shown in Fig. 6.3. Notice that the pre-emphasis makes the upper harmonics of F_0 more distinct, and the distribution of energy across the frequency range is more balanced.

6.3.2 Framing

Speech signal changes continuously due to the articulatory movements therefore the signal must be analyzed in short segments or *frames*, assuming local stationarity within each frame. The selection of the frame length is a crucial parameter for

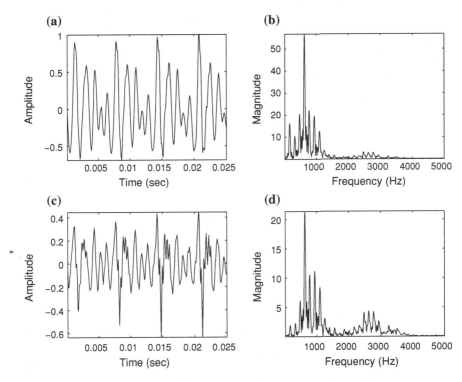

Fig. 6.3 a Time and **b** frequency domain representation of a speech signal without pre-emphasis. **c** and **d** after pre-emphasis, ($\alpha = 0.95$)

successful spectral analysis, due to the trade-off between the time and frequency resolutions. The frame should be long enough for adequate frequency resolution, but on the other hand, it should be short enough so that it would capture the local spectral properties. Typically a frame length of 10–30 milliseconds is used with an overlap of 25–50% of the frame length [47]. Overlapping is needed to avoid loss of information.

6.3.3 Windowing

The purpose of windowing is to reduce the effect of the spectral artifacts that result from the framing process [48–51]. Windowing in time domain is a point wise multiplication of the frame and the window function. A good window function has a narrow main lobe and low sidelobe levels in their transfer functions [48, 50]. There is a trade-off between these two: making the main lobe narrower increases the sidelobe levels, and vice versa. In general, a proper window function tapers smoothly down at the edges of the frame so that the effect of the discontinuities is diminished.

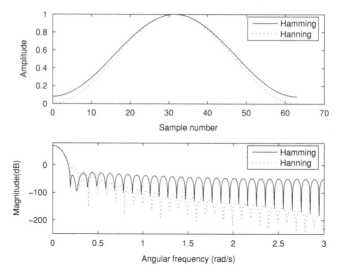

Fig. 6.4 Waveforms and magnitude response of Hamming and Hanning windows

The commonly used windows during the frequency analysis of speech sounds are Hamming and Hanning windows. They both belong to raised cosine windows family. Figure 6.4 shows the waveforms and magnitude responses of Hamming and Hanning window function. Notice that the Hamming window does not get as close to zero near the edges as the Hanning window does. In magnitude response, the main lobes of both Hamming and Hanning windows have the same width whereas the Hamming window has lower side lobes adjacent to the main lobe than the Hanning window, hence Hamming window is preferred.

6.4 Investigating Importance of Different Frequency Bands for Speaker Identification

In the past, in many speaker recognition experiments, the frequency band of the speech signal was limited up to 4 kHz assuming that most of the speaker-specific information is contained in the low frequencies. Limiting the frequencies up to 4 kHz is suitable only for speaker identification applications used over public switched telephone network (PSTN) which uses narrowband speech, nominally limited to about 200–3400 Hz and sampled at a rate of 8 kHz. However recent research have shown that a rich amount of speaker individual information is contained in the high frequency band and is useful for speaker recognition [4, 5, 52, 53]. Today, the increasing penetration of end-to-end digital networks such as the second- and third-generation wireless systems (2G and 3G) and voice over packet networks permits the use of wider speech bandwidth.

To investigate the contribution of different frequency regions for speaker identification, the following experiments were carried out. Qualitatively, these experiments show that the speaker-specific information lies not only in the range of frequencies commonly exploited for speaker and speech recognition (below 4 kHz), but extend to the higher part of the frequency scale, between 4 and 8 kHz. Similar conclusion was also drawn by Mishra et al. in [6] and Lu et al. in [52]. Table 6.1 shows the speaker identification performance for 4 different experiments. To evaluate the results, 100 speakers (64 male and 36 female from eight different dialects) from TIMIT database are considered. In TIMIT database, the SA, SX and SI sentences are recorded in the same session. The classification engine used in all these tests was based on a 32 mixtures GMM classifier.

6.4.1 Experiment 1

In the feature extraction process, speech signal is first pre-emphasized using a pre-emphasize filter, $H(z) = 1 - 0.97z^{-1}$. The pre-emphasized speech signal is then divided into 32 ms frames with 16 ms overlap. After multiplying with Hamming window, short time Fourier transform (STFT) of each frame is obtained. As the speaker individual information is not distributed uniformly in each frequency band, the mel frequency analysis is not suitable for speaker individual information extraction [52, 53]. Therefore the mel scale warping is not used while extracting the features. To obtain cepstral features, eighty triangle-shaped band pass filters with linear frequency scale were used. Each filter band gives an output which integrates the frequency energy around the center frequency of the filter band. In this experiment, the speech signal used is of 8 kHz bandwidth and the interest lies in investigating speaker-specific information in different frequency regions. Therefore to capture such information in small frequency bands, more number of filters are used. After taking logarithm of energy of each filtered signal, the DCT is applied to get 36 order cepstral coefficient vectors. Figure 6.5a shows the block schematic of the feature extractor. This experiment shows 100% correct identification (as in Table 6.1).

6.4.2 Experiment 2

In this experiment, the speaker identification rate is separately obtained for two frequency bands, 0–4 and 4–8 kHz. This experiment is carried out to quantitatively evaluate the importance of the higher part of the frequency axis (frequency range 4–8 kHz). The speech signal is divided into two bands 0–4 and 4–8 kHz by using a pair of quadrature mirror filters (QMF). The lowpass QMF coefficients, h_n are [−0.0010773, 0.0047773, 0.00055384, −0.031582, 0.027523, 0.097502, −0.12977, −0.22626, 0.31525, 0.75113, 0.49462, 0.11154]. The coefficients of the high pass filter, g_n are calculated from the h_n coefficients as,

Table 6.1 Speaker
identification performance
evaluated on different
frequency bands

Experiment	Frequency band (kHz)	Identification (%)
1	0–8	100
2	0–4	97.33
	4–8	94.66
3	0–2	54
	2–4	71.34
	4–6	69.34
	6–8	80.67
4	4–5	24
	5–6	36
	6–7	42.66
	7–8	46.67

$$g_n = (-1)^n h_{N-1-n}; \qquad N = 12. \tag{6.3}$$

Output of each filter is processed separately to obtain 36 dimensions feature vector
from each band. The feature extraction procedure is same as in Experiment 1 except
the number of filters used. The block schematic of the feature extractor is shown in
Fig. 6.5b. In this experiment, a filter bank of 40 triangle shaped linearly spaced filters
is used separately for 0–4 and 4–8 kHz bands. Let us consider a band limited signal
in the range 0–4 kHz. As maximum frequency component in this signal is 4 kHz,
ideally it can be sampled with sampling frequency of 8 kHz. With 8 kHz sampling
frequency, 32 ms frame contains only 256 samples (which are 512 in Experiment 1).
Therefore it is required to reduce the number of filters used in the filterbank. The
identification rate obtained is 97.33 and 94.66% for the frequency bands; 0–4 and
4–8 kHz respectively. It shows that both these frequency bands carry speaker-specific
information.

6.4.3 Experiment 3

Experiment 2 shows that both, 0–4 kHz as well as 4–8 kHz frequency bands are
equally important for speaker identification. Furthermore, to evaluate the importance of frequency bands with better resolution, in this experiment, the speech signal
is passed through a bank of 4 bandpass filters each with bandwidth equal to 2 kHz
with pass bands as 0–2, 2–4, 4–6 and 6–8 kHz. The filter bank is implemented using
6th order Daubechies' orthogonal filters. The output of each filter was processed
separately to obtain 18 dimensions feature vector from each frequency band. The
feature extraction procedure is same as in Experiment 2 except the number of
triangular filters used which are 20 in this case. From the identification rate given in
Table 6.1, it is clear that speaker-specific information also lies in higher frequency
bands, i.e., 4–6 and 6–8 kHz.

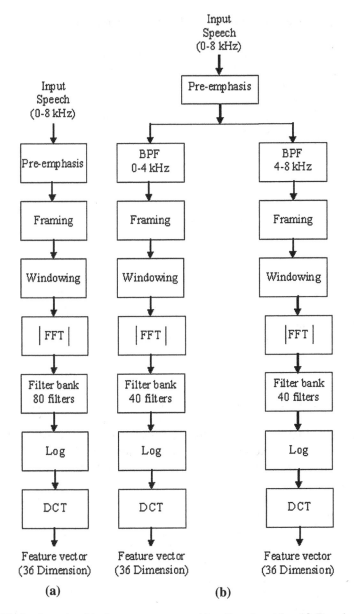

Fig. 6.5 Block schematic of the feature extractor used in **a** Experiment 1 and **b** Experiment 2

6.4.4 Experiment 4

The main aim of all above experiments is to investigate the importance of high frequency components in identifying speakers. Experiment 3 shows that 4–6 kHz as well as 6–8 kHz bands are useful for speaker discrimination. Therefore in this experiment, only the higher frequency components, in the range 4–8 kHz are considered. The resolution is further increased by reducing the filter bandwidth to 1 kHz. The identification rate is obtained separately for 4 frequency bands, 4–5, 5–6, 6–7 and 7–8 kHz. Each band pass filtered signal is further processed through a bank of 20 triangle shaped linearly spaced filters, to obtain 18 dimensions feature vector, similar to Experiment 3. The results show that the higher frequency bands 6–7 kHz as well as 7–8 kHz play an important role in speaker discrimination. Based on these results a new non-uniform filter structure is proposed for feature extraction in the following section.

6.5 Speaker Identification Using TEO

According to the results obtained in Sect. 6.4, a sub-band processing algorithm can be designed to extract features for speaker identification. There are two solutions to emphasize the contribution of the frequency bands with more speaker-specific information. The first solution is to assign weighting coefficients for different frequency bands, based on their importance in identifying speakers. The second solution is to use non-uniform frequency warping, which applies different frequency resolutions to different frequency regions according to their importance in identifying speakers. The second approach, i.e., fine tuning the bandwidth of the filters is a standard practice found in many approaches [52, 53].

Based on the second approach, a non-uniform sub-band filter bank is designed to change frequency resolutions in different frequency regions. As MFCC features are the dominant features used in most of the state-of-the-art speaker recognition systems, the Mel filter-like structure (high resolution in low frequency region and low resolution in high frequency region) is used in the 0–4 kHz frequency band. It is implemented using wavelet transform. A 6 level decomposition is applied to an interval of 0–1 kHz; a 5 level decomposition is applied to 1–2 kHz interval and a 4 level decomposition to 2–3 and 3–4 kHz intervals. Experiment 4 shows that 7–8 kHz band carries more speaker related information compared to 4–5 kHz band, and the identification rate increases gradually from 4 to 8 kHz. Based on these results, decomposition of higher frequency bands is further carried out still we get noticeable energy. Typically, 7–8 kHz band is decomposed into 8 frequency bands each of bandwidth equal to 125 Hz; 6–7 kHz band into 4 frequency bands with bandwidth of 250 Hz each; and 4–5 kHz as well as 5–6 kHz bands into 2 frequency bands with bandwidth of 500 Hz. Figure 6.6 shows the proposed filter bank structure with $L = 32$ filters. The proposed wavelet packet tree is based on the performance in high frequency range (i.e., 4–8 kHz) and approximates the mel scale in the low frequency range upto 4 kHz.

In mel scale based filter structures, generally the frequency resolution is fine in the lower frequency bands while it gets considerably coarser in the higher frequency bands. Therefore, these filter structures are not suitable to capture high frequency

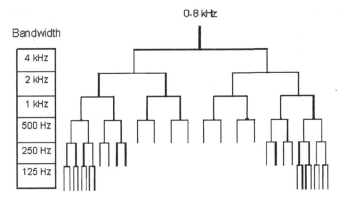

Fig. 6.6 Proposed filter bank structure

speaker-specific information. Whereas, the proposed filter structure takes the advantage of mel scale (for low frequency region, up to 4 kHz) as well as capable to capture high frequency speaker-specific information.

The input speech signal $x[n]$ is divided into 32 subsignals, $x_l[n]$, $l = 1, ..., L = 32$ using this filter structure. For every subsignal, the average Teager energy e_l is estimated as,

$$e_l = \frac{1}{N_l} \sum_{n=0}^{N_l-1} |\Psi(x_l[n])|; \qquad l = 1, ..., L \tag{6.4}$$

where N_l is the number of samples in the lth band. The magnitude of the Teager energy is used to ensure the nonnegativity of e_l, because in some rare circumstances, Teager energy has a negative value [54].

Finally, log compression and IDCT computation is applied to obtain the coefficients as,

$$TC(k) = \sum_{l=1}^{L} \log(e_l) cos \left[\frac{k(l-1/2)\pi}{L} \right]; \qquad k = 1, ..., N. \tag{6.5}$$

The idea of computing cepstrum of TEO profile of speech signal is also explored by Jankowski in [55, 56]. These coefficients are called as Teager energy operator based cepstrum coefficients (TEOCC) [57]. The first 24 TC coefficients are used in the feature vector for training and testing.

To evaluate the performance, noise from the NOISEX-92 database [58] has been added to clean speech database (TIMIT) with different SNRs. NOISEX-92 is a noise database which provides various noise signals recorded in real environments. In many real life applications, the test speech may be corrupted with noise. Car engine noise and bubble noise signals are frequently encountered in the real life. Therefore, in the experiments performed, 'car' and 'babble' noise have been used. The car noise is

Table 6.2 Speaker
identification performance in
percent for MFCC, and
TEOCC features with
different population sizes

Population	Features	
	MFCC	TEOCC
200	99	99
400	98.33	98.33
630	98.25	98.25

recorded inside a Volvo-340 on an asphalt road, in rainy conditions. In case of babble
noise, the source of the babble is 100 people speaking in a canteen. The room radius
is over two meters; therefore, individual voices are slightly audible.

In order to compare the performance of the TEOCC features, MFCC features are
used as a baseline. A filter bank with 32 triangular filters is designed according to
the mel scale warping to account the 8 kHz speech bandwidth. Then 24 dimension
feature vectors are obtained excluding the zeroth coefficient.

6.5.1 Performance Evaluation for Clean Speech

In order to evaluate the performance of the proposed TEOCC features, first clean
speech is used for training as well as testing. With clean speech conditions, the
results are obtained with MFCC and TEOCC features. The performance is evaluated
for different speaker population sizes. Table 6.2 shows the results obtained. It shows
that, the TEOCC feature performance is equally good as that of MFCC features. It
shows that, for clean speech the TEOCC features are equally acceptable as that of
the well known MFCC features.

6.5.2 Performance Evaluation for Noisy Speech: Speech Corrupted by Car Engine Noise

Car engine noise is added to the testing speech utterances to obtain the SNR of
20, 10, 5 and 0 dB. Table 6.3 shows the performance evaluated using MFCC and
TEOCC features for car engine noise. It shows that at high SNR values or with
clean speech only, MFCC features work better. When the speech signal is corrupted
by car engine noise, which is mostly low pass in nature as shown in Fig. 4.4; it
is the low frequency components of the speech which are affected the most. The
MFCC features follow the mel scale, which gives better frequency resolution at low
frequencies; hence the speaker identification accuracy is poor using MFCC features
in noisy environment. It shows that the TEOCC features perform better than the
MFCC features for noisy speech.

Table 6.3 Speaker identification performance in percent with different SNRs evaluated for MFCC and TEOCC features; test speech is corrupted by car engine noise

Feature	SNR (dB)					
	Clean	30	20	10	5	0
MFCC	99	98.75	95	63.25	42.5	25.5
TEOCC	99	98.25	96	86.5	78.5	57

Table 6.4 Speaker identification performance in percent with different SNRs evaluated for MFCC and TEOCC features; test speech is corrupted by babble noise

Feature	SNR (dB)				
	30	20	10	5	0
MFCC	99	96.25	77	49.25	18.5
TEOCC	98	96.75	83	57.5	22.75

6.5.3 Performance Evaluation for Noisy Speech: Speech Corrupted by Babble Noise

Similar experiment is performed using babble noise. Table 6.4 shows the results obtained. Compared with the results in Table 6.3, speaker identification is particularly poor for the non-stationary noise like babble noise at 0 dB SNR. At higher SNR values, the TEOCC features work equally well as that of the MFCC features and at low SNR values, the identification accuracy is better than the MFCC features. Above experiments show that the TEOCC features are robust and we can achieve better speaker identification rate in noisy environment without additional processing of the signal to remove the noise.

6.5.4 Effect of Feature Vector Dimensions

Let us see the effect of feature vector dimensions on the identification accuracy. To observe this effect, 24 as well as 32 dimensions TEOCC feature vectors are obtained. The performance evaluated using car engine noise at different SNRs is shown in Table 6.5. It shows that increasing the feature vector dimensions from 24 to 32, speaker identification performance is improved especially at low SNRs. For 0 dB SNR, the identification rate increases from 57 to 76.75%.

It shows that, TEO's filtering capability can be effectively used for noise suppression to improve the speaker identification accuracy under mismatched training and testing conditions.

Table 6.5 Identification rate (ID) as a function of TEOCC feature vector dimensions (Dim) at different SNRs

SNR (dB)	Speaker ID (%)	
	24 Dim	32 Dim
30	98.25	98.25
20	96	96.25
10	86.5	91.5
5	78.5	86.5
0	57	76.25

6.6 Speaker Identification Using AM–FM Model Based Features

The speech signal, $s[n]$ is first pre-emphasized using a pre-emphasis filter with transfer function, $H(z) = 1 - 0.97z^{-1}$. The pre-emphasized speech signal is divided into 32 ms frames with 16 ms overlap and multiplied by Hamming window. Then the multiband demodulation analysis was performed. In MDA, the filterbank plays an important role. In the experiments performed, we have used a filterbank consisting of a set of Gabor band-pass filters. Gabor filters are chosen because they are optimally compact and smooth in both the time and frequency domains. This characteristic guarantees accurate amplitude and frequency estimates in the demodulation stage [59] and reduces the incidence of ringing artifacts in the time domain [56]. Bandwidth of each individual Gabor filters within the filter-bank also plays an important role. To observe the effect of bandwidth, two different filterbanks and three different experimental set-ups are used to extract the features.

In order to evaluate the performance of the features under mismatched training and testing conditions, the GMM speaker models (with 32 mixtures) are trained using clean speech and noise is added to the test data. Car engine noise and babble noise with SNR of 20, 10, 5 and 0 dB levels are added to the testing speech utterances.

6.6.1 Set-up1: Combining Instantaneous Frequency and Amplitude

As discussed in Sect. 5.6, MDA was performed to compute the instantaneous frequency and amplitude of speech resonating signals. To perform MDA, two different filterbanks are used. The first filterbank (uniform) consists of 40 Gabor filters with uniformly spaced center frequencies and constant bandwidth of 200 Hz as shown in Fig. 6.7. The second filterbank (non-uniform) consists of 40 Gabor filters which are non-uniformly spaced and the bandwidth vary according to mel-scale, which is shown in Fig. 6.8. This filterbank is very similar to the filterbank used in conventional MFCC feature extraction technique. The only difference is, instead of using triangular filters, Gabor filters are used.

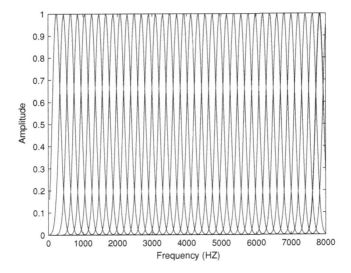

Fig. 6.7 Uniform subband filters with uniform bandwidth

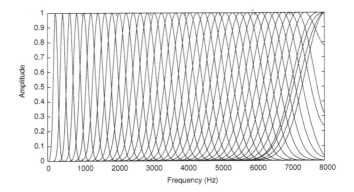

Fig. 6.8 Non-uniform subband filters with non-uniform bandwidth

After obtaining the instantaneous amplitude and frequency using HTD, the short-time mean amplitude weighted instantaneous frequency estimate is obtained using Eq. 5.12. The estimate of short-time instantaneous frequency is expressed in kilohertz in order to overcome the problem associated with the nodal variances of the GMM. Finally DCT is applied and only first 24 coefficients excluding zeroth coefficient are used to construct a feature vector. The feature vectors obtained using uniform filterbank are referred as F-1 and that of non-uniform filterbank as F-2. This feature extraction scheme is shown in Fig. 6.9.

Table 6.6 shows speaker identification rate in percent for the features obtained using set-up 1 (F-1 and F-2 features) and the MFCC features under mismatched conditions. The test speech signal is corrupted by car engine noise with different SNR

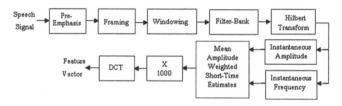

Fig. 6.9 Set-up 1: feature extraction scheme

Table 6.6 Speaker Identification performance obtained with the addition of car engine noise in the testing speech utterances at different SNRs using MFCC features and AM–FM features obtained using uniform and non-uniform filterbank, i.e., F-1 and F-2 features

Features	Speaker identification rate (%)			
	SNR = 20 dB	SNR = 10 dB	SNR = 5 dB	SNR = 0 dB
MFCC	95	63.25	42.5	25.5
F-1	86.5	77.75	70.5	57.5
F-2	97	96.75	96.5	96.25

Table 6.7 Speaker Identification performance obtained with the addition of babble noise in the testing speech utterances at different SNRs using using MFCC features and AM–FM features obtained using uniform and non-uniform filterbank, i.e., F-1 and F-2 features

Features	Speaker identification rate (%)			
	SNR = 20 dB	SNR = 10 dB	SNR = 5 dB	SNR = 0 dB
MFCC	96.25	77	49.25	18.5
F-1	89.75	79	59.25	28.5
F-2	96.75	88.25	66.5	37.75

values. The results show that speaker identification rate decreases with decreasing SNR for MFCC as well as F-1 features. Whereas the F-2 features show 96.62% average speaker identification rate irrespective of the SNR value. It shows that the F-2 features are the robust features when the noise is a car engine noise. Table 6.7 shows similar results for babble noise. It also shows that, speaker identification rate decreases with decreasing SNR. Further, speaker identification rate is particularly poor for the non-stationary noise like babble noise as compared to the car engine noise at SNR of 0 dB. At higher SNR values, the F-2 features work equally well compared to the MFCC features and at low SNR values, the identification accuracy is better than the MFCC features. It confirms that, the MFCC features are well suited only when the training and testing speech is clean (noise free) and recorded in the same environment. Furthermore, MFCC takes into account only the speech perception mechanism and not the speech production mechanism. Whereas, the features F-2, considerers both speech production (using AM–FM approach) as well as perception (non-uniform filter bank) mechanism, hence more robust compared to MFCC features.

6.6.2 Set-up 2: Combining Instantaneous Frequency and Bandwidth

In this set-up, only the non-uniform filterbank is used, because it was observed that the non-uniform filterbank shows improved results compared to uniform filterbank. After performing the demodulation using Hilbert transform, the instantaneous amplitude and frequency are combined together to obtain a mean-amplitude weighted short-time estimates, F_{iw} and B_{iw}, of the instantaneous frequency and bandwidth respectively. The estimate of short-time instantaneous frequency and bandwidth are expressed in kilohertz. Considering the basilar membrane as a bank of resonators; the quality factor 'Q' of each of these resonating filters is obtained as,

$$Q_i = \frac{F_{iw}}{B_{iw}}. \tag{6.6}$$

Finally DCT is applied and only first 24 coefficients excluding zeroth coefficient are used to construct a feature vector. These features are referred as F-3.

6.6.3 Combining Instantaneous Frequency, Bandwidth and Post Smoothing

This set-up is similar to set-up 2 except the introduction of post smoothing part. Sometimes, it is observed that, the estimates of instantaneous amplitude and frequency have singularities and spikes. To obtain robust Q features, the demodulation algorithm should provide smooth and accurate estimates. Therefore, a post-processing scheme is applied which employs a median filter with a short window (5-point). These features are referred as F-4. Here the feature extraction scheme is same as in set-up 2 with the addition of a median filter.

Table 6.8 shows speaker identification rate in percent for the features obtained using set-up 2 and set-up 3 (F-3 and F-4 features) and the MFCC features under mismatched training and testing conditions. The test speech signal is corrupted by car noise with different SNR values. It shows that F-3 as well as F-4 features are more robust in the presence of car engine noise compared to MFCC features. Table 6.9 shows the similar results for babble noise. It also shows that, F-3 and F-4 features outperform the MFCC features. While obtaining F-3 features, the quality factor of each of the filters of the non-uniform filter bank is considered. This filter bank approximately represents the basilar membrane and it is known that, the basilar membrane acts as a tuned filter bank. Further improvement can be seen in the speaker identification rate using F-4 features compared to F-3 features, for both types of noise signals. It shows that the quality factor of each of these filters can be speaker-specific and plays an important role in identifying speakers in noisy environment.

Table 6.8 Speaker Identification rate obtained with the addition of car engine noise in the testing speech utterances at different SNRs using MFCC, F-3 and F-4 features

Features	Speaker identification rate (%)			
	SNR = 20 dB	SNR = 10 dB	SNR = 5 dB	SNR = 0 dB
MFCC	95	63.25	42.5	25.5
$F-3$	95.5	95	94	93
$F-4$	98.5	98	98	97

Table 6.9 Speaker Identification rate obtained with the addition of babble noise in the testing speech utterances at different SNRs using MFCC, F-3 and F-4 features

Features	Speaker identification rate (%)			
	SNR = 20 dB	SNR = 10 dB	SNR = 5 dB	SNR = 0 dB
MFCC	96.25	77	49.25	18.5
$F-3$	93.5	82	63.5	35
$F-4$	98	92	72.5	39

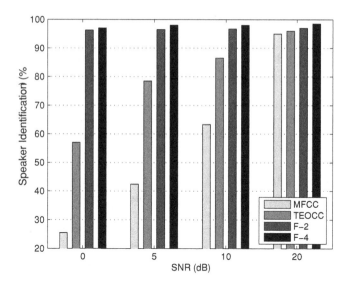

Fig. 6.10 The speaker identification performance obtained using TEOCC features, AM–FM features obtained using non-uniform filterbank (F-2), 'Q' features obtained using post smoothing technique (F-4) and the MFCC features for test speech signal corrupted by car engine noise

The features presented in this chapter can solve some of the problems of speaker identification under mismatched conditions. Figure 6.10 shows the speaker identification performance obtained using TEOCC features, AM–FM features obtained using non-uniform filterbank (F-2), 'Q' features obtained using post smoothing technique (F-4) and the MFCC features for test speech signal corrupted by car engine noise.

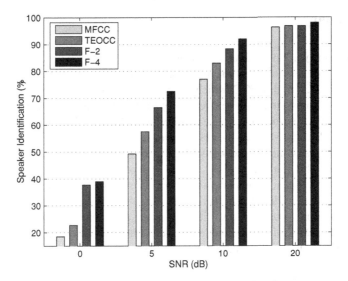

Fig. 6.11 The speaker identification performance obtained using TEOCC features, AM–FM features obtained using non-uniform filterbank (F-2), 'Q' features obtained using post smoothing technique (F-4) and the MFCC features for test speech signal corrupted by babble noise

Figure 6.11 shows the speaker identification performance obtained using TEOCC, F-2, F-4 and the MFCC features for test speech signal corrupted by babble noise. Both these graphs show that, the TEOCC and AM–FM model based features outperform the MFCC features.

MFCC features are based on the speech perception mechanism, which follow the logarithmic scale. Therefore, the filterbank used in the MFCC resolves low frequency components in better manner compared to high frequency components. However, as discussed in Sect. 6.4, the speaker-specific information lies in high frequency components also and MFCC fails to capture such information. Instead of using the conventional energy definition while extracting the features, the Teager energy shows better speaker identification rate in noisy environment because TEO provides filtering capability to suppress noise. Therefore speaker identification performance can be improved without additional processing of the noisy speech signal to suppress noise.

Further improvement in speaker identification rate can be seen by using AM–FM modeling technique. If features are derived by considering both, the speech production and speech perception mechanism, better identification rate can be achieved. Therefore the features derived from the quality factor of the resonating filters (F-2 and F-4) show significant improvement in the identification rate at low SNR. Further, it can be seen from these plots that, the proposed AM–FM based features are more robust for stationary noise like car engine noise compared to non-stationary noise (babble noise).

6.7 Summary

In this chapter, we have discussed the use of nonlinear modeling techniques for noise robust speaker identification. To overcome the limitations of the state-of-the art features like LPCC and MFCC, which are based on the linear source-filter model of speech production system, the new feature extraction techniques based on TEO and AM–FM model are proposed. These features show improved speaker identification rate compared to the state-of-the art features. It is investigated that the high frequency region also contains speaker-specific information which is useful for speaker identification. This high frequency component speaker-specific information combined with low frequency component information obtained using mel scale approach improves the speaker identification rate. However, this range is not preserved over the telephone and is not robust to noise, due to weaker speech energy at high frequencies.

Finally, the AM–FM speaker modeling is discussed. It is shown that the 'Q' factor of the resonating filters across the basilar membrane is also speaker specific. After knowing the speech production and perception mechanisms, one can derive a robust feature set based on the combination of speech production and perception systems considering their nonlinear aspects. This feature set will consider the speaker-specific cues from both, the production and perception viewpoint and hence can be more robust. It is required to investigate such parameters and some way of fusing these parameters to increase the speaker identification accuracy in noisy environments.

Now, it is well known that the dynamics of the speech signal are nonlinear and this information is useful in speech processing based applications. In the field of speech recognition, there are strong experimental evidences that acoustic features representing various aspects of the nonlinear structure of speech can increase the robustness of recognition systems. Similar evidences can also be found for applications like speech coding, speech synthesis as well as speaker recognition. Further, it can be seen that, modeling the nonlinear dynamics of the speech signal is useful in speech and speaker recognition especially under noisy conditions. However, more research is needed to simplify the nonlinear modeling techniques, feature extraction and to find optimal ways for fusing the nonlinear features with the linear speech features. To address these issues, in this book, we have tried to present the advances in nonlinear modeling techniques for speech processing with its different applications. Starting from the physical basics of the nonlinearity in human speech production, the framework for dynamic system model, Teager energy operator as well as AM–FM model has been discussed.

References

1. Prabhakar S, Pankanti S, Jain A (2003) Biometric recognition: security and privacy concerns. IEEE Secur Priv Mag 1:32–34
2. Jain AK, Ross A, Prabhakar S (2004) An introduction to biometric recognition. IEEE trans Circuits Syst Video Technol 14(1):4–20

3. Campbell JP, Shen W, Campbell WM, Schwartz R, Bonastre JF, Mastrouf D (2009) Forensic speaker recognition: a need for caution. IEEE Signal Process Mag 26(2):95–103
4. Wu JD, Lin BF (2009) Speaker identification using discrete wavelet packet transform technique with irregular decomposition. Expert Syst Appl 36:3136–3143
5. Hayakawa S, Itakura F (1994) Text-dependent speaker recognition using the information in the higher frequency band. In: Proceedings of the IEEE international conference on acoustic speech and signal processing (ICASSP'94), Adelaide, pp 137–140
6. Mishra H, Ikbal S, Yegnanarayana B (2003) Speaker specific mapping for text-independent speaker recognition. Speech Commun 39:301–310
7. Rabiner LR, Juang BH (1993) Fundamentals of speech recognition. Prentice-Hall, India
8. Patil HA, Basu TK (2004) Teager energy mel cepstrum for identification of twins in Marathi. In: IEEE India annual conference INDICON, vol 64, pp 58–61
9. Teager HM (1980) Some observations on oral air flow during phonation. IEEE Trans Speech Audio Process 28(5):599–601
10. Gish H, Schmidt M (1994) Text independent speaker identification. IEEE Signal Process Mag 11(4):18–32
11. Huggins M, Grieco J (2002) Confidence metrics for speaker identification. In: Proceedings of the international conference on spoken language processing (ICSLP'02), Denver, CO, pp 1381–1384
12. Luck JE (1969) Automatic speaker verification using cepstral measurements. J Acoust Soc Am 46(2):1026–1032
13. Pruzansky S (1963) Pattern matching procedure for automatic talker recognition. J Acoust Soc Am 35(3):354–358
14. Atal BS (1974) Effectiveness of linear prediction characteristics of the speech wave for automatic speaker identification and verification. J Acoust Soc Am 55:1304–1312
15. Sambur MR (1975) Selection of acoustic features for speaker identification. IEEE Trans Acoust Speech Signal Process 23(2):176–182
16. Rosenberg AE, Sambur MR (1975) New techniques for automatic speaker verification. IEEE Trans Acoust Speech Signal Process 23(2):169–176
17. Sambur MR (1976) Speaker recognition using orthogonal linear prediction. IEEE Trans Acoust Speech Signal Process 24(4):283–289
18. Furui S (1986) Speaker independent isolated word recognition using dynamic features of speech spectrum. IEEE Trans Acoust Speech Signal Process 34:52–59
19. Furui S (1981) Cepstral analysis technique for automatic speaker verification. IEEE Trans Acoust Speech Signal Process 29(2):254–272
20. Plumpe MD, Quatieri TF, Reynolds DA (1999) Modeling of the glottal flow derivative waveform with application to speaker identification. IEEE Trans Speech Audio Process 7(5): 569–585
21. Burton D (1987) Text-dependent speaker verification using vector quantization source coding. IEEE Trans Acoust Speech Signal Process 35(2):133–143
22. He J, Liu L, Palm G (1999) A discriminative training algorithm for VQ-based speaker identification. IEEE Trans Acoust Speech Signal Process 7(3):353–356
23. Kinnunen T, Karpov E, Franti P (2006) Real-time speaker identification and verification. IEEE Trans Audio Speech Lang Process 14(1):277–288
24. Soong F, Rosenberg A (1988) On the use of instantaneous and transitional spectral information in speaker recognition. IEEE Trans Acoust Speech Signal Process 36(6):871–879
25. Linde Y, Buzo A, Gray M (1980) An algorithm for vector quantization. IEEE Trans Commun 28(1):84–95
26. Soong F, Rosenberg A, Rabiner L, Juang B (1985) A vector quantization approach to speaker recognition. In: Proceedings of the international conference on acoustics, speech, and signal processing, vol 1, Tampa, FL, pp 387–390
27. Kinnunen T, Saastamoinen J, Hautamaki V, Vini M, Franti P (2009) Comparative evaluation of maximum a posteriori vector quantization and Gaussian mixture models in speaker verification. Pattern Recognit Lett 30(4):341–347

28. Bannani G, Gallinari P (1995) Neural networks for discrimination and modelization of speakers. Speech Commun 17:159–175
29. Yegnanarayana B (1999) Artificial neural networks. Prentice-Hall, India
30. Lipmann RP (1989) An introduction to computing with neural nets. IEEE Trans Acoust Speech Signal Process 4:4–22
31. Prasanna SRM, Gupta CS, Yegnanarayana B (2006) Extraction of speaker-specific excitation information from linear perdiction residual of speech. Speech Commun 48:1243–1261
32. Yegnanarayana B, Prasanna SRM, Zachariach JM, Gupta SC (2005) Combining evidences from source, suprasegmental and spectral features for a fixed-text speaker verification system. IEEE Trans Speech Audio Process 13(4):575–582
33. Murthy KSR, Yegnanarayana B (2006) Combining evidence from residual phase and MFCC features for speaker recognition. IEEE Signal Process Lett 13(1):52–56
34. Yegnanarayana B, Reddy KS, Kishore SP (2001) Source and system features for speaker recognition using AANN models. In: Proceedings of the IEEE international conference on acoustics, speech, and signal processing, Salt Lake city, Utah, pp 409–412
35. Reynolds DA, Rose R (1995) Robust text-independent speaker identification using Gaussian mixture speaker models. IEEE Trans Speech Audio Process 3(1):72–83
36. Reynolds DA (1995) Speaker identification and verification using Gaussian mixture speaker models. Speech Commun 17:91–108
37. Reynolds DA, Quateri TF, Dunn RB (2000) Speaker verification using adapted Gaussian mixture speaker models. Digit Signal Process 10:19–41
38. Rosenberg AE, Parthasarathy S (1996) Speaker recognition models for conected digit password speaker verification. In: Proceedings of the international conference on acoustics, speech, and signal processing (ICASSP'96), Atlanta, GA, pp 81–84
39. Matsui T, Furui S (1994) Comparison of text-independent speaker recognition methods using VQ-distortion and discrete/continuous HMMs. IEEE Trans Speech Audio Process 2(3): 456–459
40. Kimball O, Schmidt M, Gish H, Waterman J (1997) Speaker verification with limited enrollment data. In: Proceedings of the European conference on speech communication and technology (EUROSPEECH'97), Rhodes, pp 967–970
41. Deshpande MS, Holambe RS (2008) Text-independent speaker identification using hidden markov model. In: Proceedings of first IEEE international conference on emerging trends in engineering and technology (ICETET'08), Nagpur, pp 641–644
42. Wan V, Renals S (2002) Evaluation of kernel methods for speaker verification and identification. In: Proceedings of the IEEE international conference on acoustics, speech, and signal processing, vol 1, pp 669–672
43. Wan V, Renals S (2005) Speaker verification using sequence discriminant support vector machines. IEEE Trans Speech Audio Process 12:203–210
44. Campbell W, Campbell J, Reynolds D, Singer E, Torres-Carrasquillo P (2006) Support vector machines for speaker and language recognition. Comput Speech Lang 20(2):210–229
45. Campbell W, Sturim D, Reynolds D (2006) Support vector machines using GMM supervectors for speaker verification. IEEE Signal Process Lett 13(5):308–311
46. Quatieri TF (2004) Discrete-time speech signal processing principles and practice. Pearson Education, Upper Saddle River
47. Rabiner LR, Shafer RW (1989) Digital signal processing of speech signals. Prentice-Hall, Englewood Cliffs
48. Harris F (1978) On the use of windows for harmonic analysis with the discrete Fourier transform. Proc IEEE 66(1):51–84
49. Hansen J, Proakis J (2000) Discrete-time processing of speech signals, 2nd edn. IEEE Press, New York
50. Proakis J, Manolakis D (1992) Digital signal prosessing: principles, algorithms and applications, 2nd edn. Macmillan Publishing Company, New York
51. Oppenheim A, Schafer R (1975) Digital signal processing. Prentice Hall, Englewood Cliffs

52. Lu X, Dang J (2007) Physiological feature extraction for text independent speaker identification using non-uniform subband processing. In: Proceedings of the IEEE international conference on acoustic speech and signal processing (ICASSP'07), Adelaide, pp IV-461–464

53. Lu X, Dang J (2008) An investigation of dependencies between frequency components and speaker characteristics for text-independent speaker identification. Speech Commun 50: 312–322

54. Kvedalen E (2003) Signal processing using the Teager Energy Operator and other nonlinear operators. Candies Scientific Thesis, University of Oslo, Norway

55. Jankowski CR (1996) Signal processing using the Teager energy operator and other nonlinear operators. Ph.D. thesis, MIT, USA

56. Jankowski CR, Quatieri TF, Reynolds DA (1995) Measuring fine structure in speech: application to speaker identification. In: Proceedings of the IEEE international conference acoustics, speech, and, signal processing, pp 325–328

57. Jabloun F, Cetin AE, Erzin E (1999) Teager energy based feature parameters for speech recognition in car noise. IEEE Signal Process Lett 6(10):159–261

58. Noisex-92. http://www.speech.cs.cmu.edu/comp.speech/Section1/Data/noisex.html

59. Potamianos A, Maragos P (1996) Speech formant frequency and bandwidth tracking using multiband energy demodulation. J Acoust Soc Am 99(6):3795–3806

Index

A
All-pole filter, 29
AM signal, 22
AM-FM signal, 23
Amplitude envelope, 64
ANN, 4
Anti-resonances, 29
Autoregressive, 27

B
Bifurcations, 19

C
Critical band, 15

D
DESA-1, 55
DESA-1a, 55
DESA-2, 56
Dynamic system model, 35

E
Energy separation algorithm, 64
ERB, 16
ESA, 64

F
Feature extraction, 78
Fine structure, 17

FM signal, 22
Framing, 80

G
GMM, 83

H
Hilbert transform, 68
HMM, 36
HTD, 64

I
Instantaneous
    amplitude, 68
Instantaneous
    frequency, 68, 69
Instantaneous phase, 68
Instantaneous power, 45

J
Jitter, 3, 17

L
Linear model, 27
Linear source-filter
    model, 30
Log area ratios, 33
LPC, 28
LPCC, 33

R. S. Holambe and M. S. Deshpande, *Advances in Non-Linear Modeling for Speech Processing*, SpringerBriefs in Speech Technology, DOI: 10.1007/978-1-4614-1505-3, © The Author(s) 2012

M
Mel, 15
MLP, 38

N
NOISEX-92, 87
Nonlinear techniques, 11

P
Paranasal cavities, 12
Partial correlation coefficients, 33
Piecewise linear approximation, 42
Piriform fossa, 12
Pre-emphasis, 79
Prediction error, 28

Q
QMF, 85

R
Recognition, 1
Reflection coefficients, 32

S
Shimmer, 5, 18
Speaker recognition, 4, 78

Speech analysis, 3
Speech coding, 1, 3
Speech enhancement, 3
Speech perception, 14
Speech production, 12
Speech recognition, 2
Speech restoration, 4
Speech synthesis, 3, 16
State-space model, 35
Synthesis, 1

T
TEO, 45
Time-invariant linear model, 34
TIMIT, 83
Truncated taylor series, 40
Turbulent sound source, 19

V
Volterra filter, 51
Vortices, 20

W
Windowing, 81